T0273836

Biodegradable Waste and Management

Biodegradable Waste and Management

Edited by
Eliot Fox

Larsen & Keller
www.larsen-keller.com

Biodegradable Waste and Management
Edited by Eliot Fox
ISBN: 978-1-63549-040-4 (Hardback)

© 2017 Larsen & Keller

 Larsen & Keller

Published by Larsen and Keller Education,
5 Penn Plaza,
19th Floor,
New York, NY 10001, USA

Cataloging-in-Publication Data

Biodegradable waste and management / edited by Eliot Fox.
    p. cm.
Includes bibliographical references and index.
ISBN 978-1-63549-040-4
1. Refuse and refuse disposal--Biodegradation. 2. Refuse and refuse disposal. 3. Biodegradation.
4. Plastic scrap--Biodegradation. 5. Waste minimization.
I. Fox, Eliot.
TD791 .B56 2017
628.44--dc23

The publisher's policy is to use permanent paper from mills that operate a sustainable forestry policy. Furthermore, the publisher ensures that the text paper and cover boards used have met acceptable environmental accreditation standards.

Printed and bound in the United States of America.

For more information regarding Larsen and Keller Education and its products, please visit the publisher's website www.larsen-keller.com

Table of Contents

Preface

Biodegradation refers to the process of decomposition of materials by microorganisms like fungi, bacteria, etc. It is a natural process, which enables the disintegration of waste in nature. It is often used in fields like waste management, ecology, natural environment and biomedicine, etc. This book is a compilation of chapters that discuss the most vital concepts in the field of biodegradation. It is compiled in such a manner, that it will provide in-depth knowledge about the theory and practice of the subject. Most of the topics introduced in this book cover new techniques and applications of biodegradable waste management. The text aims to serve as a resource guide for students and contribute to the growth of the discipline.

To facilitate a deeper understanding of the contents of this book a short introduction of every chapter is written below:

Chapter 1- Biodegradation is a process of the break down or decomposition of organic matter by the action of microorganisms. It is vital for sustaining life on Earth. This chapter will provide an integrated understanding on biodegradation.

Chapter 2- Organic matter that can be broken down by microorganisms over a period of time are termed as biodegradable waste. Composting is a method of manually encouraging waste decomposition. Many organisms feed on this waste. This chapter educates the reader on processes such as aerobics digestion, photo degradation, decomposition and some other concepts.

Chapter 3- Plastic that can be decomposed by bacteria is known as biodegradable plastic. Plastic, by nature, can damage the environment and biodegradable plastic is an answer and a cure for that. This chapter introduces the reader to fascinating concepts like bioplastic, polycaprolactone, polyvinyl alcohol and polyproylene carbonate.

Chapter 4- This chapter focuses on different practices of disposing waste. Some of these practices are landfill, incineration and compost. Waste materials have certain organic substances. The burning of these organic substances is known as incineration. Landfills on the other hand are locations, which are used for the removal of waste materials by burial.

Chapter 5- Methods and techniques are an important component of any field of study. The following chapter elucidates the various techniques that are related to biodegradable waste management. Promession, bioremediation, pyrolysis, rotating biological contactor are some of the methods that are covered in this section.

Chapter 6- The uses stated in this chapter are biodegradable athletic footwear, biodegradable bag, dry animal dung fuel and cook stove. The content strategically encompasses and incorporates the major uses of biodegradable minerals, providing a complete understanding.

Chapter 7- Persistent organic pollutants (POPs) are substances like insecticides and pesticides that do not decompose easily or are not naturally eliminated.

The major categories of persistent organic pollutants are dealt in great detail. The aspects elucidated are of vital importance, and provide a better understanding of persistent organic pollutants.

I would like to share the credit of this book with my editorial team who worked tirelessly on this book. I owe the completion of this book to the never-ending support of my family, who supported me throughout the project.

Editor

Introduction to Biodegradation

Biodegradation is a process of the break down or decomposition of organic matter by the action of microorganisms. It is vital for sustaining life on Earth. This chapter will provide an integrated understanding on biodegradation.

Biodegradation

Yellow slime mold growing on a bin of wet paper

IUPAC Definition

Degradation caused by enzymatic process resulting from the action of cells.

Biodegradation is the disintegration of materials by bacteria, fungi, or other biological means. Although often conflated, biodegradable is distinct in meaning from compostable. While biodegradable simply means to be consumed by microorganisms, "compostable" makes the specific demand that the object break down under composting conditions. The term is often used in relation to ecology, waste management, biomedicine, and the natural environment (bioremediation) and is now commonly associated with environmentally friendly products that are capable of decomposing back into natural elements. Organic material can be degraded aerobically with oxygen, or anaerobically, without oxygen. Biosurfactant, an extracellular surfactant secreted by microorganisms, enhances the biodegradation process.

Biodegradable matter is generally organic material that serves as a nutrient for microorganisms. Microorganisms are so numerous and diverse that, a huge range of compounds are biodegraded, including hydrocarbons (e.g. oil), polychlorinated biphenyls (PCBs), polyaromatic hydrocarbons (PAHs), pharmaceutical substances. Decomposition of biodegradable substances may include both biological and abiotic steps.

Factors Affecting Rate

In practice, almost all chemical compounds and materials are subject to biodegradation, the key is the relative rates of such processes - minutes, days, years, centuries... A number of factors determine the degradation rate of organic compounds. Salient factors include light, water and oxygen. Temperature is also important because chemical reactions proceed more quickly at higher temperatures. The degradation rate of many organic compounds is limited by their bioavailability. Compounds must be released into solution before organisms can degrade them.

Biodegradability can be measured in a number of ways. Respirometry tests can be used for aerobic microbes. First one places a solid waste sample in a container with microorganisms and soil, and then aerate the mixture. Over the course of several days, microorganisms digest the sample bit by bit and produce carbon dioxide – the resulting amount of CO_2 serves as an indicator of degradation. Biodegradability can also be measured by anaerobic microbes and the amount of methane or alloy that they are able to produce. In formal scientific literature, the process is termed bio-remediation.

Approximated time for compounds to biodegrade in a marine environment	
Product	Time to Biodegrade
Paper towel	2–4 weeks
Newspaper	6 weeks
Apple core	2 months
Cardboard box	2 months
Wax coated milk carton	3 months
Cotton gloves	1–5 months
Wool gloves	1 year
Plywood	1–3 years
Painted wooden sticks	13 years
Plastic bags	10–20 years
Tin cans	50 years
Disposable diapers	50–100 years
Plastic bottle	100 years
Aluminium cans	200 years
Glass bottles	Undetermined

Detergents

In advanced societies, laundry detergents are based on *linear* alkylbenzenesulfonates. Branched alkybenzenesulfonates (below right), used in former times, were abandoned because they biodegrade too slowly.

4-(5-Dodecyl) benzenesulfonate, a linear dodecylbenzenesulfonate

A branched dodecylbenzenesulfonate, which has been phased out in developed countries.

Plastics

Plastics biodegrade at highly variable rates. PVC-based plumbing is specifically selected for handing sewage because PVC biodegrades very slowly. Some packaging materials on the other hand are being developed that would degrade readily upon exposure to the environment. Illustrative synthetic polymers that are biodegrade quickly include polycaprolactone, others are polyesters and aromatic-aliphatic esters, due to their ester bonds being susceptible to attack by water. A prominent example is poly-3-hydroxybutyrate, the renewably derived polylactic acid, and the synthetic polycaprolactone. Others are the cellulose-based cellulose acetate and celluloid (cellulose nitrate).

Polylactic acid is an example of a plastic that biodegrades quickly.

Under low oxygen conditions biodegradable plastics break down slower and with the production of methane, like other organic materials do. The breakdown process is accelerated in a dedicated compost heap. Starch-based plastics will degrade within two to four months in a home compost bin, while polylactic acid is largely undecomposed, requiring higher temperatures. Polycaprolactone and polycaprolactone-starch composites decompose slower, but the starch content accelerates decomposition by leaving behind a porous, high surface area polycaprolactone. Nevertheless, it takes many months. In 2016, a bacterium named Ideonella sakaiensis was found to biodegrade PET.

Many plastic producers have gone so far even to say that their plastics are compostable, typically listing corn starch as an ingredient. However, these claims are questionable because the plastics industry operates under its own definition of compostable:

"that which is capable of undergoing biological decomposition in a compost site such that the material is not visually distinguishable and breaks down into carbon dioxide, water, inorganic compounds and biomass at a rate consistent with known compostable materials." (Ref: ASTM D 6002)

The term "composting" is often used informally to describe the biodegradation of packaging materials. Legal definitions exist for compostability, the process that leads to compost. Four criteria are offered by the European Union:

- Biodegradability, the conversion of >90% material material into CO_2 and water by the action of micro-organisms within 6 months.

- Disintegrability, the fragmentation of 90% of the original mass to particles that then pass through a 2 mm sieve.

- Absence of toxic substances and other substances that impede composting.

Biodegradable Technology

In 1973 it was proven for the first time that polyester degrades when disposed in bioactive material such as soil. Polyesters are water resistant and can be melted and shaped into sheets, bottles, and other products, making certain plastics now available as a biodegradable product. Following, Polyhydroxylalkanoates (PHAs) were produced directly from renewable resources by microbes. They are approximately 95% cellular bacteria and can be manipulated by genetic strategies. The composition and biodegradability of PHAs can be regulated by blending it with other natural polymers. In the 1980s the company ICI Zenecca commercialized PHAs under the name Biopol. It was used for the production of shampoo bottles and other cosmetic products. Consumer response was unusual. Consumers were willing to pay more for this product because it was natural and biodegradable, which had not occurred before.

Now biodegradable technology is a highly developed market with applications in product packaging, production and medicine. Biodegradable technology is concerned with the manufacturing science of biodegradable materials. It imposes science based mechanisms of plant genetics into the processes of today. Scientists and manufacturing corporations can help impact climate change by developing a use of plant genetics that would mimic some technologies. By looking to plants, such as biodegradable material harvested through photosynthesis, waste and toxins can be minimized.

Oxo-biodegradable technology, which has further developed biodegradable plastics, has also emerged. Oxo-biodegradation is defined by CEN (the European Standards Organisation) as "degradation resulting from oxidative and cell-mediated phenomena, either simultaneously or successively." Whilst sometimes described as "oxo-fragmentable," and "oxo-degradable" this describes only the first or oxidative phase. These descriptions should not be used for material which degrades by the process of oxo-biodegradation defined by CEN, and the correct description is "oxo-biodegradable."

By combining plastic products with very large polymer molecules, which contain only carbon and hydrogen, with oxygen in the air, the product is rendered capable of decomposing in anywhere from a week to one to two years. This reaction occurs even without prodegradant additives but at a very slow rate. That is why conventional plastics, when discarded, persist for a long time in the environment. Oxo-biodegradable formulations catalyze and accelerate the biodegradation process but it takes considerable skill and experience to balance the ingredients within the formulations so as to provide the product with a useful life for a set period, followed by degradation and biodegradation.

Biodegradable technology is especially utilized by the bio-medical community. Biodegradable polymers are classified into three groups: medical, ecological, and dual application, while in terms of origin they are divided into two groups: natural and synthetic. The Clean Technology Group is exploiting the use of supercritical carbon dioxide, which under high pressure at room temperature is a solvent that can use biodegradable plastics to make polymer drug coatings. The polymer (meaning a material composed of molecules with repeating structural units that form a long chain) is used to encapsulate a drug prior to injection in the body and is based on lactic acid, a compound normally produced in the body, and is thus able to be excreted naturally. The coating is designed for controlled release over a period of time, reducing the number of injections required and maximizing the therapeutic benefit. Professor Steve Howdle states that biodegradable polymers are particularly attractive for use in drug delivery, as once introduced into the body they require no retrieval or further manipulation and are degraded into soluble, non-toxic by-products. Different polymers degrade at different rates within the body and therefore polymer selection can be tailored to achieve desired release rates.

Other biomedical applications include the use of biodegradable, elastic shape-memory polymers. Biodegradable implant materials can now be used for minimally invasive surgical procedures through degradable thermoplastic polymers. These polymers are now able to change their shape with increase of temperature, causing shape memory capabilities as well as easily degradable sutures. As a result, implants can now fit through small incisions, doctors can easily perform complex deformations, and sutures and other material aides can naturally biodegrade after a completed surgery.

Etymology of "Biodegradable"

The first known use of the word in biological text was in 1961 when employed to describe the breakdown of material into the base components of carbon, hydrogen, and oxygen by microorganisms. Now biodegradable is commonly associated with environmentally friendly products that are part of the earth's innate cycle and capable of decomposing back into natural elements.

Oxo Biodegradable

OXO Biodegradable OXO-biodegradation is defined by CEN (the European Standards Organisation) {CEN/TR 1535-2006} as "degradation resulting from oxidative and cell-mediated phenomena, either simultaneously or successively." Sometimes described as "OXO-degradable" this describes only the first or oxidative phase. These descriptions should not be used for material which degrades by the process of OXO-biodegradation defined by CEN, and the correct description is "OXO-biodegradable."

There are two different types of biodegradable plastic:

1. Oxo-biodegradable plastic, made from polymers such as PE (polyethylene), PP (polypropylene), and PS (polystyrene) containing extra ingredients (not heavy metals) and tested according to ASTM D6954 or BS8472 or AFNOR Accord T51-808 to degrade and biodegrade in the open environment

2. Vegetable based plastics (also loosely knows as bio-plastics "bioplastics" or "compostable plastics"). These are tested in accordance with ASTM D6400 or EN13432 to biodegrade in the conditions found in industrial composting or biogas facilities.

OXO-bio plastic is conventional polyolefin plastic to which has been added small amounts of metal salts, none of which are "heavy metals" which are restricted by the EU Packaging Waste Directive 94/62 Art 11. These salts catalyze the degradation process to speed it up so that the OXO plastic will degrade abiotically at the end of its useful life in the presence of oxygen much more quickly than ordinary plastic. At the end of that process it is no longer visible, it is no longer a plastic as it has been converted via Carboxylation or Hydroxylation to small-chain organic chemicals which will then biodegrade. It does not therefore leave fragments of plastic in the environment. The degradation process is shortened from decades to years and/or months for abiotic degradation and thereafter the rate of biodegradation depends on the micro-organisms in the environment. It does not however need to be in a highly microbial environment such as compost. Timescale for complete biodegradation is much shorter than for "conventional" plastics which, in normal environments, are very slow to biodegrade and cause large scale harm.

The useful life of a product made using oxo-biodegradable plastic can be programmed at manufacture, typically 6 months for a bread wrapper and 18 months for a lighweight, plastic carrier bag to allow for re-use. Oxo-biodegradable plastic can be manufactured with the existing machinery and workforce in factories at little or no extra cost. They have the same strength and other characteristics as ordinary plastics during their intended lifetime.

Degradation Process

Degradation is a process that takes place in many materials. The speed depends on the environment. Conventional polyethylene (PE) and polypropylene (PP) plastics will typically take decades to degrade. But OXO-biodegradable products utilize a prodegradant to speed up the molecular breakdown of the polyolefins and to incorporate oxygen atoms into the resulting low molecular mass molecules. This chemical change enables the further breakdown of the material by naturally-occurring micro-organisms.

The first process of degradation in OXO-treated plastic is an oxidative chain scission that is catalyzed by metal salts leading to oxygenated (hydroxylated and carboxylated) shorter-chain molecules .

OXO plastic, if discarded in the environment, will degrade to oxygenated low molecular weight chains (typically MW 5-10 000 amu) within 2–18 months depending on the material (resin, thickness, anti-oxidants, etc.) and the temperature and other factors in the environment.

Illustration of the OXO-Degradation: A process whereby the conventional polyolefin plastic is first oxidised to short-chain oxygenated molecules which are biodegradable (typically after two to four months of exposure)

OXO plastics are designed so that they will not degrade deep in landfill and they will not therefore generate methane (a powerful greenhouse gas) in anaerobic conditions.

OXO-biodegradable products do not degrade immediately in an open environment because they are stabilized to give the product a useful service-life. They will nevertheless degrade and biodegrade in nature if they are exposed to the environment as litter much quicker than natural waste such as twigs and straw and much more quickly than ordinary plastic. OXO-bio plastics will degrade indoors, but this is not their purpose. They are intended to degrade and biodegrade by a synergistic process in the open environment.

OXO-biodegradation of polymer material has been studied in depth at the Technical Research Institute of Sweden and the Swedish University of Agricultural Sciences. A peer-reviewed report of the work was published in Vol 96 of the journal of Polymer Degradation & Stability (2011) at page 919-928. It shows 91% biodegradation in a soil environment within 24 months, when tested in accordance with ISO 17556.

Standards Applicability

OXO-biodegradable plastic degrades in the presence of oxygen, and the process is accelerated by UV and heat. It can be recycled during its useful life with normal plastic. It is not designed to be compostable in industrial composting facilities according to ASTM D6400 or EN13432, but it can be composted in an in-vessel process.

These standards require the material to convert to CO_2 gas within 180 days because industrial composting has a short timescale and is not the same as degradation in the open environment. A leaf is generally considered to be biodegradable but it will not pass the composting standards due to the 180-day limit. (Indeed, materials which do comply with AST D6400, EN13432, Australian

4736 and ISO 17088 cannot properly be described as "compostable." This is because those standards require them to convert substantially to CO_2 gas within 180 days. You cannot therefore make them into compost - only into CO_2 gas. This contributes to climate change, but does nothing for the soil.

There is an American Standard (ASTM D6954) and a British Standard (BS8472) which specifies procedures to test degradability, biodegradability, and non-toxicity, and with which a properly designed and manufactured OXO product complies. It also contains pass/fail criteria to exclude any significant gel content which might inhibit degradation.

There is no need to refer to a Standard Specification unless a specific disposal route (e.g.: composting), is envisaged. ASTM D6400 Australian 4736 and EN13432 are Standard Specifications appropriate only for the special conditions found in industrial composting.

Another reference document has recently been published by the French standards organisation AFNOR. This document AC.51 808 offers a well researched method to test OXO-biodegradable plastics based on usage and climate conditions. It introduces a new testing method for the biodegradation of polymer using selected microorganisms and measuring ATP and ADP by chemiluminescence. This method brings a new approach as tests are done at 37 °C which is much more relevant to outdoor conditions than ASTM D6400 or EN 13432 done at 58 °C plus the microorganisms are identified based on the environment in which the plastic is likely to be disposed, which is not the case with the CO_2-evolution method.

This French document is a very interesting innovation for predicting the behaviour of an OXO-biodegradable plastic in case of littering. This test method provides an ecotoxicity testing method to ensure that residues in the environment, pending complete biodegradation, are not toxic for the Rhodococcus rhodochrous ATCC 29672 bacterium strain.

Environmental Issues

OXO-bio plastics, especially in the form of plastic bags, are now used in many places as a solution to the problem of plastic litter in the open environment. They are mandatory in some areas of the Middle-East, Asia and Africa. OXO-biodegradable plastics contain metal salts used as the catalyst for OXO-biodegradation. These carry no risks of environmental pollution as they do not contain heavy metals but rather transition metals such as iron, cobalt, manganese or nickel. A Life-cycle assessment made by INTERTEK in May 2012 classified OXO as the leading solution to address the problem of littering.

OXO-bio products have to pass the eco-toxicity tests in ASTM D6954; they are designed not to degrade deep in landfill so that they will not generate methane. There is no evidence of any danger to wildlife; almost all the plastic fragments found in studies on the marine environment are fragments of conventional plastic, unsurprisingly as this still makes up the vast majority of plastics in circulation; some argue OXO-biodegradability removes the main rationale for bans on plastic-bag i.e. that conventional plastic can lie or float around in the environment for decades.

The other rationale is that oil-reserves should not be used to make plastic, but oil is extracted primarily to make fuels, and plastic is made as an inevitable by-product of the refining process. Some argue that the same amount of oil would therefore be extracted if plastic did not exist. Another issue often discussed is whether OXO-bio plastic can be safely recycled with other oil-based plastics. The Roediger report, commissioned by the European Plastic Converters trade association, of 5 December 2013 found that it can, and that most bio-based "compostable" plastic cannot.

There is no evidence that OXO-biodegradable plastic of any kind encourages littering or discourages recycling, and it is in fact indistinguishable to the naked eye from conventional plastic. Assessing the Environmental Impacts of Oxo-biodegradable Plastics Across Their Life Cycle]). The stored energy potential of OXO-biodegradable plastic could be retrieved by thermal recycling if collected during its useful life, as it has a similar calorific value to the raw product (fossil fuel) from which it was made.

It is clear that millions of tons of plastics are in the environment and a lot of countries do not have the capacity to recycle them. It is estimated that of the 300 million tons of plastic produced annually in the world only 3% is recycled. OXO offers an option for plastic in the environment which cannot realistically be collected, but should not be treated as a reason not to collect plastic if possible.

Microbial Biodegradation

Microbial biodegredation is the use of bioremediation and biotransformation methods to harness the naturally occurring ability of microbial xenobiotic metabolism to degrade, transform or accumulate environmental pollutants, including hydrocarbons (e.g. oil), polychlorinated biphenyls (PCBs), polyaromatic hydrocarbons (PAHs), heterocyclic compounds (such as pyridine or quinoline), pharmaceutical substances, radionuclides and metals.

Interest in the microbial biodegradation of pollutants has intensified in recent years, and recent major methodological breakthroughs have enabled detailed genomic, metagenomic, proteomic, bioinformatic and other high-throughput analyses of environmentally relevant microorganisms, providing new insights into biodegradative pathways and the ability of organisms to adapt to changing environmental conditions.

Biological processes play a major role in the removal of contaminants and take advantage of the catabolic versatility of microorganisms to degrade or convert such compounds. In environmental microbiology, genome-based global studies are increasing the understanding of metabolic and regulatory networks, as well as providing new information on the evolution of degradation pathways and molecular adaptation strategies to changing environmental conditions.

Aerobic Biodegradation of Pollutants

The increasing amount of bacterial genomic data provides new opportunities for understanding the genetic and molecular bases of the degradation of organic pollutants. Aromatic compounds are among the most persistent of these pollutants and lessons can be learned from the recent genomic studies of *Burkholderia xenovorans* LB400 and *Rhodococcus* sp. strain RHA1, two of the largest bacterial genomes completely sequenced to date. These studies have helped expand our understanding of bacterial catabolism, non-catabolic physiological adaptation to organic compounds, and the evolution of large bacterial genomes. First, the metabolic pathways from phylogenetically diverse isolates are very similar with respect to overall organization. Thus, as originally noted in pseudomonads, a large number of "peripheral aromatic" pathways funnel a range of natural and xenobiotic compounds into a restricted number of "central aromatic" pathways. Nevertheless, these pathways are genetically organized in genus-specific fashions, as exemplified by the b-ketoadipate and Paa pathways. Comparative genomic studies further reveal that some pathways are more widespread than initially thought. Thus, the Box and Paa pathways illustrate the prevalence of non-oxygenolytic ring-cleavage strategies in aerobic aromatic degradation processes. Functional genomic studies have been useful in establishing that even organisms harboring high numbers of homologous enzymes seem to contain few examples of true redundancy. For example, the multiplicity of ring-cleaving dioxygenases in certain rhodococcal isolates may be attributed to the cryptic aromatic catabolism of different terpenoids and steroids. Finally, analyses have indicated that recent genetic flux appears to have played a more significant role in the evolution of some large genomes, such as LB400's, than others. However, the emerging trend is that the large gene repertoires of potent pollutant degraders such as LB400 and RHA1 have evolved principally through more ancient processes. That this is true in such phylogenetically diverse species is remarkable and further suggests the ancient origin of this catabolic capacity.

Anaerobic Biodegradation of Pollutants

Anaerobic microbial mineralization of recalcitrant organic pollutants is of great environmental significance and involves intriguing novel biochemical reactions. In particular, hydrocarbons and halogenated compounds have long been doubted to be degradable in the absence of oxygen, but the isolation of hitherto unknown anaerobic hydrocarbon-degrading and reductively dehalogenating bacteria during the last decades provided ultimate proof for these processes in nature. While such research involved mostly chlorinated compounds initially, recent studies have revealed reductive dehalogenation of bromine and iodine moieties in aromatic pesticides. Other reactions, such as biologically induced abiotic reduction by soil minerals, has been shown to deactivate relatively persistent aniline-based herbicides far more rapidly than observed in aerobic environments. Many novel biochemical reactions were discovered enabling the respective metabolic pathways, but progress in the molecular understanding of these bacteria was rather slow, since genetic systems are not readily applicable for most of them. However, with the increasing application of genomics in the field of environmental microbiology, a new and promising perspective is now at hand to obtain molecular insights into these new metabolic properties. Several complete genome sequences were determined during the last few years from bacteria capable of anaerobic organic pollutant degradation. The ~4.7 Mb genome of the facultative denitrifying *Aromatoleum aromaticum* strain EbN1 was the first to be determined for an anaerobic hydrocarbon degrader (using toluene or ethylbenzene as substrates). The genome sequence revealed about two dozen gene clusters (including several paralogs) coding for a complex catabolic network for anaerobic and aerobic degradation of aromatic compounds. The genome sequence forms the basis for current detailed studies on regulation of pathways and enzyme structures. Further genomes of anaerobic hydrocarbon degrading bacteria were recently completed for the iron-reducing species *Geobacter metallireducens* (accession nr. NC_007517) and the perchlorate-reducing *Dechloromonas aromatica* (accession nr. NC_007298), but these are not yet evaluated in formal publications. Complete genomes were also determined for bacteria capable of anaerobic degradation of halogenated hydrocarbons by halorespiration: the ~1.4 Mb genomes of *Dehalococcoides ethenogenes* strain 195 and *Dehalococcoides* sp. strain CBDB1 and the ~5.7 Mb genome of *Desulfitobacterium hafniense* strain Y51. Characteristic for all these bacteria is the presence of multiple paralogous genes for reductive dehalogenases, implicating a wider dehalogenating spectrum of the organisms than previously known. Moreover, genome sequences provided unprecedented insights into the evolution of reductive dehalogenation and differing strategies for niche adaptation.

Recently, it has become apparent that some organisms, including *Desulfitobacterium chlororespirans*, originally evaluated for halorespiration on chlorophenols, can also use certain brominated compounds, such as the herbicide bromoxynil and its major metabolite as electron acceptors for growth. Iodinated compounds may be dehalogenated as well, though the process may not satisfy the need for an electron acceptor.

Bioavailability, Chemotaxis, and Transport of Pollutants

Bioavailability, or the amount of a substance that is physiochemically accessible to microorganisms is a key factor in the efficient biodegradation of pollutants. O'Loughlin *et al.* (2000) showed that, with the exception of kaolinite clay, most soil clays and cation exchange resins attenuated biodegradation of 2-picoline by *Arthrobacter* sp. strain R1, as a result of adsorption of the substrate to the clays. Chemotaxis, or the directed movement of motile organisms towards or away from chemicals in the environment is an important physiological response that may contribute to effective catabolism of molecules in the environment. In addition, mechanisms for the intracellular accumulation of aromatic molecules via various transport mechanisms are also important.

Oil Biodegradation

Petroleum oil contains aromatic compounds that are toxic to most life forms. Episodic and chronic pollution of the environment by oil causes major disruption to the local ecological environment. Marine environments in particular are especially vulnerable, as oil spills near coastal regions and in the open sea are difficult to contain and make mitigation efforts more complicated. In addition to pollution through human activities, approximately 250 million litres of petroleum enter the marine environment every year from natural seepages. Despite its toxicity, a considerable fraction of petroleum oil entering marine systems is eliminated by the hydrocarbon-degrading activities of microbial communities, in particular by a recently discovered group of specialists, the hydrocarbonoclastic bacteria (HCB). *Alcanivorax borkumensis* was the first HCB to have its genome sequenced. In addition to hydrocarbons, crude oil often contains various heterocyclic compounds, such as pyridine, which appear to be degraded by similar mechanisms to hydrocarbons.

Cholesterol Biodegradation

Many synthetic steroidic compounds like some sexual hormones frequently appear in municipal and industrial wastewaters, acting as environmental pollutants with strong metabolic activities negatively affecting the ecosystems. Since these compounds are common carbon sources for many different microorganisms their aerobic and anaerobic mineralization has been extensively studied. The interest of these studies lies on the biotechnological applications of sterol transforming enzymes for the industrial synthesis of sexual hormones and corticoids. Very recently, the catabolism of cholesterol has acquired a high relevance because it is involved in the infectivity of the pathogen

Mycobacterium tuberculosis (*Mtb*). *Mtb* causes tuberculosis disease, and it has been demonstrated that novel enzyme architectures have evolved to bind and modify steroid compounds like cholesterol in this organism and other steroid-utilizing bacteria as well. These new enzymes might be of interest for their potential in the chemical modification of steroid substrates.

Analysis of Waste Biotreatment

Sustainable development requires the promotion of environmental management and a constant search for new technologies to treat vast quantities of wastes generated by increasing anthropogenic activities. Biotreatment, the processing of wastes using living organisms, is an environmentally friendly, relatively simple and cost-effective alternative to physico-chemical clean-up options. Confined environments, such as bioreactors, have been engineered to overcome the physical, chemical and biological limiting factors of biotreatment processes in highly controlled systems. The great versatility in the design of confined environments allows the treatment of a wide range of wastes under optimized conditions. To perform a correct assessment, it is necessary to consider various microorganisms having a variety of genomes and expressed transcripts and proteins. A great number of analyses are often required. Using traditional genomic techniques, such assessments are limited and time-consuming. However, several high-throughput techniques originally developed for medical studies can be applied to assess biotreatment in confined environments.

Metabolic Engineering and Biocatalytic Applications

The study of the fate of persistent organic chemicals in the environment has revealed a large reservoir of enzymatic reactions with a large potential in preparative organic synthesis, which has already been exploited for a number of oxygenases on pilot and even on industrial scale. Novel catalysts can be obtained from metagenomic libraries and DNA sequence based approaches. Our increasing capabilities in adapting the catalysts to specific reactions and process requirements by rational and random mutagenesis broadens the scope for application in the fine chemical industry, but also in the field of biodegradation. In many cases, these catalysts need to be exploited in whole cell bioconversions or in fermentations, calling for system-wide approaches to understanding strain physiology and metabolism and rational approaches to the engineering of whole cells as they are increasingly put forward in the area of systems biotechnology and synthetic biology.

Fungal Biodegradation

In the ecosystem, different substrates are attacked at different rates by consortia of organisms from different kingdoms. *Aspergillus* and other moulds play an important role in these consortia because they are adept at recycling starches, hemicelluloses, celluloses, pectins and other sugar polymers. Some aspergilli are capable of degrading more refractory compounds such as fats, oils, chitin, and keratin. Maximum decomposition occurs when there is sufficient nitrogen, phosphorus and other essential inorganic nutrients. Fungi also provide food for many soil organisms.

For *Aspergillus* the process of degradation is the means of obtaining nutrients. When these moulds degrade human-made substrates, the process usually is called biodeterioration. Both paper and textiles (cotton, jute, and linen) are particularly vulnerable to *Aspergillus* degradation. Our artistic heritage is also subject to *Aspergillus* assault. To give but one example, after Florence in Italy flooded in 1969, 74% of the isolates from a damaged Ghirlandaio fresco in the Ognissanti church were *Aspergillus versicolor*.

References

- Koukkou, Anna-Irini, ed. (2011). Microbial Bioremediation of Non-metals: Current Research. Caister Academic Press. ISBN 978-1-904455-83-7.

- Díaz, Eduardo, ed. (2008). Microbial Biodegradation: Genomics and Molecular Biology (1st ed.). Caister Academic Press. ISBN 978-1-904455-17-2.

- McLeod MP & Eltis LD (2008). "Genomic Insights Into the Aerobic Pathways for Degradation of Organic Pollutants". Microbial Biodegradation: Genomics and Molecular Biology. Caister Academic Press. ISBN 978-1-904455-17-2.

- Heider J & Rabus R (2008). "Genomic Insights in the Anaerobic Biodegradation of Organic Pollutants". Microbial Biodegradation: Genomics and Molecular Biology. Caister Academic Press. ISBN 978-1-904455-17-2.

- Martins dos Santos VA, et al. (2008). "Genomic Insights into Oil Biodegradation in Marine Systems". In Díaz E. Microbial Biodegradation: Genomics and Molecular Biology. Caister Academic Press. ISBN 978-1-904455-17-2.

- Watanabe K & Kasai Y (2008). "Emerging Technologies to Analyze Natural Attenuation and Bioremediation". Microbial Biodegradation: Genomics and Molecular Biology. Caister Academic Press. ISBN 978-1-904455-17-2.

- Meyer A & Panke S (2008). "Genomics in Metabolic Engineering and Biocatalytic Applications of the Pollutant Degradation Machinery". Microbial Biodegradation: Genomics and Molecular Biology. Caister Academic Press. ISBN 978-1-904455-17-2.

- Machida, Masayuki; Gomi, Katsuya, eds. (2010). Aspergillus: Molecular Biology and Genomics. Caister Academic Press. ISBN 978-1-904455-53-0.

- Bennett JW (2010). "An Overview of the Genus Aspergillus" (PDF). Aspergillus: Molecular Biology and Genomics. Caister Academic Press. ISBN 978-1-904455-53-0.

- Heider J & Rabus R (2008). "Genomic Insights in the Anaerobic Biodegradation of Organic Pollutants". Microbial Biodegradation: Genomics and Molecular Biology. Caister Academic Press. ISBN 978-1-904455-17-2.

- Parales RE, et al. (2008). "Bioavailability, Chemotaxis, and Transport of Organic Pollutants". Microbial Biodegradation: Genomics and Molecular Biology. Caister Academic Press. ISBN 978-1-904455-17-2.

- "Microsoft Word - SECTION 6 BIODEGRADABILITY OF PACKAGING WASTE.doc" (PDF). Www3.imperial.ac.uk. Retrieved 2014-03-02.

- Wipperman, Matthew, F.; Sampson, Nicole, S.; Thomas, Suzanne, T. (2014). "Pathogen roid rage: Cholesterol utilization by Mycobacterium tuberculosis". Crit Rev Biochem Mol Biol. 49 (4): 269–93. doi:10.3109/10409238.2014.895700. PMID 24611808.

- Thomas, S.T.; Sampson, N.S. (2013). "Mycobacterium tuberculosis utilizes a unique heterotetrameric structure for dehydrogenation of the cholesterol side chain". Biochemistry. 52 (17): 2895–2904. doi:10.1021/bi4002979. PMC 3726044. PMID 23560677.

A Comprehensive Study of Biodegradable Waste and Related Concepts

Organic matter that can be broken down by microorganisms over a period of time are termed as biodegradable waste. Composting is a method of manually encouraging waste decomposition. Many organisms feed on this waste. This chapter educates the reader on processes such as aerobics digestion, photo degradation, decomposition and some other concepts.

Biodegradable Waste

Biodegradable waste includes any organic matter in waste which can be broken down into carbon dioxide, water, methane or simple organic molecules by micro-organisms and other living things using composting, aerobic digestion, anaerobic digestion or similar processes. In waste management, it also includes some inorganic materials which can be decomposed by bacteria. Such materials include gypsum and its products such as plasterboard and other simple organic sulfates which can be decomposed to yield hydrogen sulfide in anaerobic land-fill conditions.

In domestic waste collection, the scope of biodegradable waste may be narrowed to include only those degradable wastes capable of being handled in the local waste handling facilities.

Sources

Biodegradable waste can be commonly found in municipal solid waste (sometimes called biodegradable municipal waste, or BMW) as green waste, food waste, paper waste, and biodegradable plastics. Other biodegradable wastes include human waste, manure, sewage, sewage sludge and slaughterhouse waste. In the absence of oxygen, much of this waste will decay to methane by anaerobic digestion.

In many parts of the developed world, biodegradable waste is separated from the rest of the waste stream, either by separate kerb-side collection or by waste sorting after collection. At the point of collection such waste is often referred to as *Green waste*. Removing such waste from the rest of the waste stream substantially reduces waste volumes for disposal and also allows biodegradable waste to be composted where composting facilities exist.

Uses of Biodegradable Waste

Biodegradable waste can be used for composting or a resource for heat, electricity and fuel by means of incineration or anaerobic digestion. Swiss *Kompogas* and the Danish *AIKAN* process are examples of anaerobic digestion of biodegradable waste. While incineration can recover the most energy, anaerobic digestion plants retain the nutrients and compost for the soil and still recover some of the contained energy in the form of biogas. Kompogas produced 27 million Kwh of electricity and biogas in 2009. The oldest of the company's own lorries has achieved 1,000,000 kilometers driven with biogas from household waste in the last 15 years.

Areas Relying on Organic Waste

Featured in an edition of *The Economist* that predicted events in 2014, it was revealed that Massachusetts creates roughly 1.4 million tons of organic waste every year. Massachusetts, along with Connecticut and Vermont, are also going to enact laws to divert food waste from landfills.

In small and densely populated states, landfill capacity is limited so disposal costs are higher ($60–90 per ton in MA compared to national average of $45). Decomposing food waste generates methane, a notorious greenhouse gas. However, this biogas can be captured and turned into energy through anaerobic digestion, and then sold into the electricity grid.

Anaerobic digestion grew in Europe, but is starting to develop in America. Massachusetts is increasing its production of anaerobic digesters.

Climate Change Impacts

The main environmental threat from biodegradable waste is the production of methane and other greenhouse gases.

Aerobic Digestion

Aerobic digestion is a process in sewage treatment designed to reduce the volume of sewage sludge and make it stable appropriate for subsequent use. It is a bacterial process occurring in the presence of oxygen. Bacteria rapidly consume organic matter and convert it into carbon dioxide, water and a range of lower molecular weight organic compounds. As there is no new supply of organic material from sewage, the activated sludge biota begin to die and are used as food by saporotrophic bacteria. This stage of the process is known as *endogenous respiration* and it is process that reduces the solid concentration in the sludge.

Process

Aerobic digestion is typically used in an activated sludge treatment plant. Waste activated sludge and primary sludge are combined, where appropriate, and passed to a thickener where the solids content is increased. This substantially reduces the volume that is required to be treated in the digester.The process is usually run as a batch process with more than one digester tank in operation at any one time. Air is pumped through the tank and the contents are stirred to keep the contents fully mixed. Carbon dioxide, waste air and small quantities of other gases including hydrogen sulfide are given off. These waste gases require treatment to reduce odours in works close to housing or capable of generating public nuisance. The digestion is continued until the percentage is degradable solids is reduced to between 20% and 10% depending on local conditions

Advantages

Because the aerobic digestion occurs much faster than anaerobic digestion, the capital costs of aerobic digestion are lower.

The process is usually run at ambient temperature and the process is much less complex than anaerobic digestion and is easier to manage.

Disadvantages

The operating costs are typically much greater for aerobic digestion than for anaerobic digestion because of energy used by the blowers, pumps and motors needed to add oxygen to the process. However, recent technological advances include non-electrically aerated filter systems that use natural air currents for the aeration instead of electrically operated machinery.

The digested sludge is relatively low in residual energy and although it can be dried and incinerated to produce heat, the energy yield is very much lower than that produced by anaerobic digestion

Anaerobic Digestion

Anaerobic digestion is a collection of processes by which microorganisms break down biodegradable material in the absence of oxygen. The process is used for industrial or domestic purposes to manage waste and/or to produce fuels. Much of the fermentation used industrially to produce food and drink products, as well as home fermentation, uses anaerobic digestion.

Anaerobic digestion occurs naturally in some soils and in lake and oceanic basin sediments, where it is usually referred to as "anaerobic activity". This is the source of marsh gas methane as discovered by Volta in 1776.

The digestion process begins with bacterial hydrolysis of the input materials. Insoluble organic polymers, such as carbohydrates, are broken down to soluble derivatives that become available for other bacteria. Acidogenic bacteria then convert the sugars and amino acids into carbon dioxide, hydrogen, ammonia, and organic acids. These bacteria convert these resulting organic acids into acetic acid, along with additional ammonia, hydrogen, and carbon dioxide. Finally, methanogens convert these products to methane and carbon dioxide. The methanogenic archaea populations play an indispensable role in anaerobic wastewater treatments.

It is used as part of the process to treat biodegradable waste and sewage sludge. As part of an integrated waste management system, anaerobic digestion reduces the emission of landfill gas into the atmosphere. Anaerobic digesters can also be fed with purpose-grown energy crops, such as maize.

Anaerobic digestion is widely used as a source of renewable energy. The process produces a biogas, consisting of methane, carbon dioxide and traces of other 'contaminant' gases. This biogas can be used directly as fuel, in combined heat and power gas engines or upgraded to natural gas-quality biomethane. The nutrient-rich digestate also produced can be used as fertilizer.

With the re-use of waste as a resource and new technological approaches which have lowered capital costs, anaerobic digestion has in recent years received increased attention among governments in a number of countries, among these the United Kingdom (2011), Germany and Denmark (2011).

Process

Many microorganisms affect anaerobic digestion, including acetic acid-forming bacteria (acetogens) and methane-forming archaea (methanogens). These organisms promote a number of chemical processes in converting the biomass to biogas.

Gaseous oxygen is excluded from the reactions by physical containment. Anaerobes utilize electron acceptors from sources other than oxygen gas. These acceptors can be the organic material itself or may be supplied by inorganic oxides from within the input material. When the oxygen source in an anaerobic system is derived from the organic material itself, the 'intermediate' end products are primarily alcohols, aldehydes, and organic acids, plus carbon dioxide. In the presence of specialised methanogens, the intermediates are converted to the 'final' end products of methane, carbon dioxide, and trace levels of hydrogen sulfide. In an anaerobic system, the majority of the chemical energy contained within the starting material is released by methanogenic bacteria as methane.

Populations of anaerobic microorganisms typically take a significant period of time to establish themselves to be fully effective. Therefore, common practice is to introduce anaerobic microorganisms from materials with existing populations, a process known as "seeding" the digesters, typically accomplished with the addition of sewage sludge or cattle slurry.

Process Stages

The four key stages of anaerobic digestion involve hydrolysis, acidogenesis, acetogenesis and methanogenesis. The overall process can be described by the chemical reaction, where organic material such as glucose is biochemically digested into carbon dioxide (CO_2) and methane (CH_4) by the anaerobic microorganisms.

$$C_6H_{12}O_6 \rightarrow 3CO_2 + 3CH_4$$

- Hydrolysis

In most cases, biomass is made up of large organic polymers. For the bacteria in anaerobic digesters to access the energy potential of the material, these chains must first be broken down into their smaller constituent parts. These constituent parts, or monomers, such as sugars, are readily available to other bacteria. The process of breaking these chains and dissolving the smaller molecules into solution is called hydrolysis. Therefore, hydrolysis of these high-molecular-weight polymeric components is the necessary first step in anaerobic digestion. Through hydrolysis the complex organic molecules are broken down into simple sugars, amino acids, and fatty acids.

Acetate and hydrogen produced in the first stages can be used directly by methanogens. Other molecules, such as volatile fatty acids (VFAs) with a chain length greater than that of acetate must first be catabolised into compounds that can be directly used by methanogens.

- Acidogenesis

The biological process of acidogenesis results in further breakdown of the remaining components by acidogenic (fermentative) bacteria. Here, VFAs are created, along with ammonia, carbon dioxide, and hydrogen sulfide, as well as other byproducts. The process of acidogenesis is similar to the way milk sours.

- Acetogenesis

The third stage of anaerobic digestion is acetogenesis. Here, simple molecules created through the acidogenesis phase are further digested by acetogens to produce largely acetic acid, as well as carbon dioxide and hydrogen.

- Methanogenesis

The terminal stage of anaerobic digestion is the biological process of methanogenesis. Here, methanogens use the intermediate products of the preceding stages and convert them into methane, carbon dioxide, and water. These components make up the majority of the biogas emitted from the system. Methanogenesis is sensitive to both high and low pHs and occurs between pH 6.5 and pH 8. The remaining, indigestible material the microbes cannot use and any dead bacterial remains constitute the digestate.

Configuration

Anaerobic digesters can be designed and engineered to operate using a number of different configurations and can be categorized into batch vs. continuous process mode, mesophilic vs. thermophilic temperature conditions, high vs. low portion of solids, and single stage vs. multistage processes. More initial build money and a larger volume of the batch digester is needed to handle the same amount of waste as a continuous process digester. Higher heat energy is demanded in a thermophilic system compared to a mesophilic system and has a larger gas output capacity and higher methane gas content. For solids content, low will handle up to 15% solid content. Above this level is considered high solids content and can also be known as dry digestion. In a single stage process, one reactor houses the four anaerobic digestion steps. A multistage process utilizes two or more reactors for digestion to separate the methanogenesis and hydrolysis phases.

Batch or Continuous

Anaerobic digestion can be performed as a batch process or a continuous process. In a batch system, biomass is added to the reactor at the start of the process. The reactor is then sealed for the duration of the process. In its simplest form batch processing needs inoculation with already processed material to start the anaerobic digestion. In a typical scenario, biogas production will be formed with a normal distribution pattern over time. Operators can use this fact to determine when they believe the process of digestion of the organic matter has completed. There can be severe odour issues if a batch reactor is opened and emptied before the process is well completed. A more advanced type of batch approach has limited the odour issues by integrating anaerobic digestion with in-vessel composting. In this approach inoculation takes place through the use of recirculated degasified percolate. After anaerobic digestion has completed, the biomass is kept in the reactor which is then used for in-vessel composting before it is opened As the batch digestion is simple and requires less equipment and lower levels of design work, it is typically a cheaper form of digestion. Using more than one batch reactor at a plant can ensure constant production of biogas.

In continuous digestion processes, organic matter is constantly added (continuous complete mixed) or added in stages to the reactor (continuous plug flow; first in – first out). Here, the end products are constantly or periodically removed, resulting in constant production of biogas. A single or multiple digesters in sequence may be used. Examples of this form of anaerobic digestion include continuous stirred-tank reactors, upflow anaerobic sludge blankets, expanded granular sludge beds and internal circulation reactors.

Temperature

The two conventional operational temperature levels for anaerobic digesters determine the species of methanogens in the digesters:

- *Mesophilic* digestion takes place optimally around 30 to 38 °C, or at ambient temperatures between 20 and 45 °C, where mesophiles are the primary microorganism present.

- *Thermophilic* digestion takes place optimally around 49 to 57 °C, or at elevated temperatures up to 70 °C, where thermophiles are the primary microorganisms present.

A limit case has been reached in Bolivia, with anaerobic digestion in temperature working conditions of less than 10 °C. The anaerobic process is very slow, taking more than three times the normal mesophilic time process. In experimental work at University of Alaska Fairbanks, a 1,000 litre digester using psychrophiles harvested from "mud from a frozen lake in Alaska" has produced 200–300 litres of methane per day, about 20 to 30% of the output from digesters in warmer climates. Mesophilic species outnumber thermophiles, and they are also more tolerant to changes in environmental conditions than thermophiles. Mesophilic systems are, therefore, considered to be more stable than thermophilic digestion systems. In contrast, while thermophilic digestion systems are considered less stable, their energy input is higher, with more biogas being removed from the organic matter in an equal amount of time. The increased temperatures facilitate faster reaction rates, and thus faster gas yields. Operation at higher temperatures facilitates greater pathogen reduction of the digestate. In countries where legislation, such as the Animal By-Products Regulations in the European Union, requires digestate to meet certain levels of pathogen reduction there may be a benefit to using thermophilic temperatures instead of mesophilic.

Additional pre-treatment can be used to reduce the necessary retention time to produce biogas. For example, certain processes shred the substrates to increase the surface area or use a thermal pretreatment stage (such as pasteurisation) to significantly enhance the biogas output. The pasteurisation process can also be used to reduce the pathogenic concentration in the digesate leaving the anaerobic digester. Pasteurisation may be achieved by heat treatment combined with maceration of the solids.

Solids Content

In a typical scenario, three different operational parameters are associated with the solids content of the feedstock to the digesters:

- High solids (dry—stackable substrate)

- High solids (wet—pumpable substrate)

- Low solids (wet—pumpable substrate)

High solids (dry) digesters are designed to process materials with a solids content between 25 and 40%. Unlike wet digesters that process pumpable slurries, high solids (dry – stackable substrate) digesters are designed to process solid substrates without the addition of water. The primary styles of dry digesters are continuous vertical plug flow and batch tunnel horizontal digesters. Continuous vertical plug flow digesters are upright, cylindrical tanks where feedstock is continuously fed into the top of the digester, and flows downward by gravity during digestion. In batch tunnel digesters, the feedstock is deposited in tunnel-like chambers with a gas-tight door. Neither approach has mixing inside the digester. The amount of pretreatment, such as contaminant removal, depends both upon the nature of the waste streams being processed and the desired quality of the digestate. Size reduction (grinding) is beneficial in continuous vertical systems, as it accelerates digestion, while batch systems avoid grinding and instead require structure (e.g. yard waste) to reduce compaction of the stacked pile. Continuous vertical dry digesters have a smaller footprint due to the shorter effective retention time and vertical design. Wet digesters can be designed to operate in either a high-solids content, with a total suspended solids (TSS) concentration greater than ~20%, or a low-solids concentration less than ~15%.

High solids (wet) digesters process a thick slurry that requires more energy input to move and process the feedstock. The thickness of the material may also lead to associated problems with abrasion. High solids digesters will typically have a lower land requirement due to the lower volumes associated with the moisture. High solids digesters also require correction of conventional performance calculations (e.g. gas production, retention time, kinetics, etc.) originally based on very dilute sewage digestion concepts, since larger fractions of the feedstock mass are potentially convertible to biogas.

Low solids (wet) digesters can transport material through the system using standard pumps that require significantly lower energy input. Low solids digesters require a larger amount of land than high solids due to the increased volumes associated with the increased liquid-to-feedstock ratio of the digesters. There are benefits associated with operation in a liquid environment, as it enables more thorough circulation of materials and contact between the bacteria and their food. This enables the bacteria to more readily access the substances on which they are feeding, and increases the rate of gas production.

Complexity

Digestion systems can be configured with different levels of complexity. In a single-stage digestion system (one-stage), all of the biological reactions occur within a single, sealed reactor or holding tank. Using a single stage reduces construction costs, but results in less control of the reactions occurring within the system. Acidogenic bacteria, through the production of acids, reduce the pH of the tank. Methanogenic bacteria, as outlined earlier, operate in a strictly defined pH range. Therefore, the biological reactions of the different species in a single-stage reactor can be in direct competition with each other. Another one-stage reaction system is an anaerobic lagoon. These lagoons are pond-like, earthen basins used for the treatment and long-term storage of manures. Here the anaerobic reactions are contained within the natural anaerobic sludge contained in the pool.

In a two-stage digestion system (multistage), different digestion vessels are optimised to bring maximum control over the bacterial communities living within the digesters. Acidogenic bacteria produce organic acids and more quickly grow and reproduce than methanogenic bacteria. Methanogenic bacteria require stable pH and temperature to optimise their performance.

Under typical circumstances, hydrolysis, acetogenesis, and acidogenesis occur within the first reaction vessel. The organic material is then heated to the required operational temperature (either mesophilic or thermophilic) prior to being pumped into a methanogenic reactor. The initial hydrolysis or acidogenesis tanks prior to the methanogenic reactor can provide a buffer to the rate at which feedstock is added. Some European countries require a degree of elevated heat treatment to kill harmful bacteria in the input waste. In this instance, there may be a pasteurisation or sterilisation stage prior to digestion or between the two digestion tanks. Notably, it is not possible to completely isolate the different reaction phases, and often some biogas is produced in the hydrolysis or acidogenesis tanks.

Residence Time

The residence time in a digester varies with the amount and type of feed material, and with the configuration of the digestion system. In a typical two-stage mesophilic digestion, residence time varies between 15 and 40 days, while for a single-stage thermophilic digestion, residence times is normally faster and takes around 14 days. The plug-flow nature of some of these systems will mean the full degradation of the material may not have been realised in this timescale. In this event, digestate exiting the system will be darker in colour and will typically have more odour.

In the case of an upflow anaerobic sludge blanket digestion (UASB), hydraulic residence times can be as short as 1 hour to 1 day, and solid retention times can be up to 90 days. In this manner, a UASB system is able to separate solids and hydraulic retention times with the use of a sludge blanket. Continuous digesters have mechanical or hydraulic devices, depending on the level of solids in the material, to mix the contents, enabling the bacteria and the food to be in contact. They also allow excess material to be continuously extracted to maintain a reasonably constant volume within the digestion tanks.

Inhibition

Left: Farm-based maize silage digester located near Neumünster in Germany, 2007 - the green, inflatable biogas holder is shown on top of the digester. *Right:* Two-stage, low solids, UASB digestion component of a mechanical biological treatment system near Tel Aviv; the process water is seen in balance tank and sequencing batch reactor, 2005.

Feedstocks

The most important initial issue when considering the application of anaerobic digestion systems is the feedstock to the process. Almost any organic material can be processed with anaerobic digestion; however, if biogas production is the aim, the level of putrescibility is the key factor in its successful application. The more putrescible (digestible) the material, the higher the gas yields possible from the system.

Anaerobic lagoon and generators at the Cal Poly Dairy, United States

Feedstocks can include biodegradable waste materials, such as waste paper, grass clippings, left-over food, sewage, and animal waste. Woody wastes are the exception, because they are largely unaffected by digestion, as most anaerobes are unable to degrade lignin. Xylophalgeous anaerobes (lignin consumers) or using high temperature pretreatment, such as pyrolysis, can be used to break down the lignin. Anaerobic digesters can also be fed with specially grown energy crops, such as silage, for dedicated biogas production. In Germany and continental Europe, these facilities are referred to as "biogas" plants. A codigestion or cofermentation plant is typically an agricultural anaerobic digester that accepts two or more input materials for simultaneous digestion.

The length of time required for anaerobic digestion depends on the chemical complexity of the material. Material rich in easily digestible sugars breaks down quickly where as intact lignocellulosic material rich in cellulose and hemicellulose polymers can take much longer to break down. Anaerobic microorganisms are generally unable to break down lignin, the recalcitrant aromatic component of biomass.

Anaerobic digesters were originally designed for operation using sewage sludge and manures. Sewage and manure are not, however, the material with the most potential for anaerobic digestion, as the biodegradable material has already had much of the energy content taken out by the animals that produced it. Therefore, many digesters operate with codigestion of two or more types of feedstock. For example, in a farm-based digester that uses dairy manure as the primary feedstock, the gas production may be significantly increased by adding a second feedstock, e.g., grass and corn (typical on-farm feedstock), or various organic byproducts, such as slaughterhouse waste, fats, oils and grease from restaurants, organic household waste, etc. (typical off-site feedstock).

Digesters processing dedicated energy crops can achieve high levels of degradation and biogas production. Slurry-only systems are generally cheaper, but generate far less energy than those using crops, such as maize and grass silage; by using a modest amount of crop material (30%), an anaerobic digestion plant can increase energy output tenfold for only three times the capital cost, relative to a slurry-only system.

Moisture Content

A second consideration related to the feedstock is moisture content. Dryer, stackable substrates, such as food and yard waste, are suitable for digestion in tunnel-like chambers. Tunnel-style systems typically have near-zero wastewater discharge, as well, so this style of system has advantages where the discharge of digester liquids are a liability. The wetter the material, the more suitable it will be to handling with standard pumps instead of energy-intensive concrete pumps and physical means of movement. Also, the wetter the material, the more volume and area it takes up relative to the levels of gas produced. The moisture content of the target feedstock will also affect what type of system is applied to its treatment. To use a high-solids anaerobic digester for dilute feedstocks, bulking agents, such as compost, should be applied to increase the solids content of the input material. Another key consideration is the carbon:nitrogen ratio of the input material. This ratio is the balance of food a microbe requires to grow; the optimal C:N ratio is 20–30:1. Excess N can lead to ammonia inhibition of digestion.

Contamination

The level of contamination of the feedstock material is a key consideration. If the feedstock to the digesters has significant levels of physical contaminants, such as plastic, glass, or metals, then processing to remove the contaminants will be required for the material to be used. If it is not removed, then the digesters can be blocked and will not function efficiently. It is with this understanding that mechanical biological treatment plants are designed. The higher the level of pretreatment a feedstock requires, the more processing machinery will be required, and, hence, the project will have higher capital costs.

After sorting or screening to remove any physical contaminants from the feedstock, the material is often shredded, minced, and mechanically or hydraulically pulped to increase the surface area available to microbes in the digesters and, hence, increase the speed of digestion. The maceration of solids can be achieved by using a chopper pump to transfer the feedstock material into the airtight digester, where anaerobic treatment takes place.

Substrate Composition

Substrate composition is a major factor in determining the methane yield and methane production rates from the digestion of biomass. Techniques to determine the compositional characteristics of the feedstock are available, while parameters such as solids, elemental, and organic analyses are important for digester design and operation.

Applications

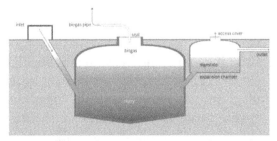

Schematic of an anaerobic digester as part of a sanitation system. It produces a digested slurry (digestate) that can be used as a fertilizer, and biogas that can be used for energy.

Using anaerobic digestion technologies can help to reduce the emission of greenhouse gases in a number of key ways:

- Replacement of fossil fuels
- Reducing or eliminating the energy footprint of waste treatment plants
- Reducing methane emission from landfills
- Displacing industrially produced chemical fertilizers
- Reducing vehicle movements
- Reducing electrical grid transportation losses
- Reducing usage of LP Gas for cooking

Waste and Wastewater Treatment

Anaerobic digestion is particularly suited to organic material, and is commonly used for industrial effluent, wastewater and sewage sludge treatment. Anaerobic digestion, a simple process, can greatly reduce the amount of organic matter which might otherwise be destined to be dumped at sea, dumped in landfills, or burnt in incinerators.

Pressure from environmentally related legislation on solid waste disposal methods in developed countries has increased the application of anaerobic digestion as a process for reducing waste volumes and generating useful byproducts. It may either be used to process the source-separated fraction of municipal waste or alternatively combined with mechanical sorting systems, to process residual mixed municipal waste. These facilities are called mechanical biological treatment plants.

If the putrescible waste processed in anaerobic digesters were disposed of in a landfill, it would break down naturally and often anaerobically. In this case, the gas will eventually escape into the atmosphere. As methane is about 20 times more potent as a greenhouse gas than carbon dioxide, this has significant negative environmental effects.

In countries that collect household waste, the use of local anaerobic digestion facilities can help to reduce the amount of waste that requires transportation to centralized landfill sites or incineration facilities. This reduced burden on transportation reduces carbon emissions from the collection vehicles. If localized anaerobic digestion facilities are embedded within an electrical distribution network, they can help reduce the electrical losses associated with transporting electricity over a national grid.

Power Generation

In developing countries, simple home and farm-based anaerobic digestion systems offer the potential for low-cost energy for cooking and lighting. From 1975, China and India have both had large, government-backed schemes for adaptation of small biogas plants for use in the household for cooking and lighting. At present, projects for anaerobic digestion in the developing world can gain financial support through the United Nations Clean Development Mechanism if they are able to show they provide reduced carbon emissions.

Methane and power produced in anaerobic digestion facilities can be used to replace energy derived from fossil fuels, and hence reduce emissions of greenhouse gases, because the carbon in biodegradable material is part of a carbon cycle. The carbon released into the atmosphere from the combustion of biogas has been removed by plants for them to grow in the recent past, usually within the last decade, but more typically within the last growing season. If the plants are regrown, taking the carbon out of the atmosphere once more, the system will be carbon neutral. In contrast, carbon in fossil fuels has been sequestered in the earth for many millions of years, the combustion of which increases the overall levels of carbon dioxide in the atmosphere.

Biogas from sewage works is sometimes used to run a gas engine to produce electrical power, some or all of which can be used to run the sewage works. Some waste heat from the engine is then used to heat the digester. The waste heat is, in general, enough to heat the digester to the required temperatures. The power potential from sewage works is limited – in the UK, there are about 80 MW total of such generation, with the potential to increase to 150 MW, which is insignificant compared to the average power demand in the UK of about 35,000 MW. The scope for biogas generation from nonsewage waste biological matter – energy crops, food waste, abattoir waste, etc. - is much higher, estimated to be capable of about 3,000 MW. Farm biogas plants using animal waste and energy crops are expected to contribute to reducing CO_2 emissions and strengthen the grid, while providing UK farmers with additional revenues.

Some countries offer incentives in the form of, for example, feed-in tariffs for feeding electricity onto the power grid to subsidize green energy production.

In Oakland, California at the East Bay Municipal Utility District's main wastewater treatment plant (EBMUD), food waste is currently codigested with primary and secondary municipal wastewater solids and other high-strength wastes. Compared to municipal wastewater solids digestion alone, food waste codigestion has many benefits. Anaerobic digestion of food waste pulp from the EBMUD food waste process provides a higher normalized energy benefit, compared to municipal wastewater solids: 730 to 1,300 kWh per dry ton of food waste applied compared to 560 to 940 kWh per dry ton of municipal wastewater solids applied.

Grid Injection

Biogas grid-injection is the injection of biogas into the natural gas grid. The raw biogas has to be previously upgraded to biomethane. This upgrading implies the removal of contaminants such as hydrogen sulphide or siloxanes, as well as the carbon dioxide. Several technologies are available for this purpose, being the most widely implemented the pressure swing adsorption (PSA), water or amine scrubbing (absorption processes) and, in the last years, membrane separation. As an alternative, the electricity and the heat can be used for on-site generation, resulting in a reduction of losses in the transportation of energy. Typical energy losses in natural gas transmission systems range from 1–2%, whereas the current energy losses on a large electrical system range from 5–8%.

In October 2010, Didcot Sewage Works became the first in the UK to produce biomethane gas supplied to the national grid, for use in up to 200 homes in Oxfordshire. By 2017, UK electricity firm Ecotricity plan to have digester fed by locally sourced grass fueling 6000 homes

Vehicle Fuel

After upgrading with the above-mentioned technologies, the biogas (transformed into biomethane) can be used as vehicle fuel in adapted vehicles. This use is very extensive in Sweden, where over 38,600 gas vehicles exist, and 60% of the vehicle gas is biomethane generated in anaerobic digestion plants.

Fertiliser and Soil Conditioner

The solid, fibrous component of the digested material can be used as a soil conditioner to increase the organic content of soils. Digester liquor can be used as a fertiliser to supply vital nutrients to soils instead of chemical fertilisers that require large amounts of energy to produce and transport. The use of manufactured fertilisers is, therefore, more carbon-intensive than the use of anaerobic digester liquor fertiliser. In countries such as Spain, where many soils are organically depleted, the markets for the digested solids can be equally as important as the biogas.

Cooking Gas

By using a bio-digester, which produces the bacteria required for decomposing, cooking gas is generated. The organic garbage like fallen leaves, kitchen waste, food waste etc. are fed into a crusher unit, where the mixture is conflated with a small amount of water. The mixture is then fed into the bio-digester, where the bacteria decomposes it to produce cooking gas. This gas is piped to kitchen stove. A 2 cubic meter bio-digester can produce 2 cubic meter of cooking gas. This is equivalent to 1 kg of LPG. The notable advantage of using a bio-digester is the sludge which is a rich organic manure.

Products

The three principal products of anaerobic digestion are biogas, digestate, and water.

Biogas

Biogas is the ultimate waste product of the bacteria feeding off the input biodegradable feedstock (the methanogenesis stage of anaerobic digestion is performed by archaea (a micro-organism on a distinctly different branch of the phylogenetic tree of life to bacteria), and is mostly methane and carbon dioxide, with a small amount hydrogen and trace hydrogen sulfide. (As-produced, biogas also contains water vapor, with the fractional water vapor volume a function of biogas temperature). Most of the biogas is produced during the middle of the digestion, after the bacterial population has grown, and tapers off as the putrescible material is exhausted. The gas is normally stored on top of the digester in an inflatable gas bubble or extracted and stored next to the facility in a gas holder.

Typical Composition of Biogas		
Compound	Formula	%
Methane	CH_4	50–75
Carbon dioxide	CO_2	25–50
Nitrogen	N_2	0–10
Hydrogen	H_2	0–1
Hydrogen sulphide	H_2S	0–3
Oxygen	O_2	0–0

The methane in biogas can be burned to produce both heat and electricity, usually with a reciprocating engine or microturbine often in a cogeneration arrangement where the electricity and waste heat generated are used to warm the digesters or to heat buildings. Excess electricity can be sold to suppliers or put into the local grid. Electricity produced by anaerobic digesters is considered to be renewable energy and may attract subsidies. Biogas does not contribute to increasing atmospheric carbon dioxide concentrations because the gas is not released directly into the atmosphere and the carbon dioxide comes from an organic source with a short carbon cycle.

Biogas may require treatment or 'scrubbing' to refine it for use as a fuel. Hydrogen sulfide, a toxic product formed from sulfates in the feedstock, is released as a trace component of the biogas. National environmental enforcement agencies, such as the U.S. Environmental Protection Agency or the English and Welsh Environment Agency, put strict limits on the levels of gases containing hydrogen sulfide, and, if the levels of hydrogen sulfide in the gas are high, gas scrubbing and cleaning equipment (such as amine gas treating) will be needed to process the biogas to within regionally accepted levels. Alternatively, the addition of ferrous chloride $FeCl_2$ to the digestion tanks inhibits hydrogen sulfide production.

Volatile siloxanes can also contaminate the biogas; such compounds are frequently found in household waste and wastewater. In digestion facilities accepting these materials as a component of the feedstock, low-molecular-weight siloxanes volatilise into biogas. When this gas is combusted in a gas engine, turbine, or boiler, siloxanes are converted into silicon dioxide ($SiO2$), which deposits internally in the machine, increasing wear and tear. Practical and cost-effective technologies to remove siloxanes and other biogas contaminants are available at the present time. In certain applications, *in situ* treatment can be used to increase the methane purity by reducing the offgas carbon dioxide content, purging the majority of it in a secondary reactor.

Biogas holder with lightning protection rods and backup gas flare

In countries such as Switzerland, Germany, and Sweden, the methane in the biogas may be compressed for it to be used as a vehicle transportation fuel or input directly into the gas mains. In countries where the driver for the use of anaerobic digestion are renewable electricity subsidies, this route of treatment is less likely, as energy is required in this processing stage and reduces the overall levels available to sell.

Biogas carrying pipes

Digestate

Digestate is the solid remnants of the original input material to the digesters that the microbes cannot use. It also consists of the mineralised remains of the dead bacteria from within the digesters. Digestate can come in three forms: fibrous, liquor, or a sludge-based combination of the two fractions. In two-stage systems, different forms of digestate come from different digestion tanks. In single-stage digestion systems, the two fractions will be combined and, if desired, separated by further processing.

Acidogenic anaerobic digestate

The second byproduct (acidogenic digestate) is a stable, organic material consisting largely of lignin and cellulose, but also of a variety of mineral components in a matrix of dead bacterial cells; some plastic may be present. The material resembles domestic compost and can be used as such or to make low-grade building products, such as fibreboard. The solid digestate can also be used as feedstock for ethanol production.

The third byproduct is a liquid (methanogenic digestate) rich in nutrients, which can be used as a fertiliser, depending on the quality of the material being digested. Levels of potentially toxic elements (PTEs) should be chemically assessed. This will depend upon the quality of the original feedstock. In the case of most clean and source-separated biodegradable waste streams, the levels of PTEs will be low. In the case of wastes originating from industry, the levels of PTEs may be higher and will need to be taken into consideration when determining a suitable end use for the material.

Digestate typically contains elements, such as lignin, that cannot be broken down by the anaerobic microorganisms. Also, the digestate may contain ammonia that is phytotoxic, and may hamper the growth of plants if it is used as a soil-improving material. For these two reasons, a maturation or composting stage may be employed after digestion. Lignin and other materials are available for degradation by aerobic microorganisms, such as fungi, helping reduce the overall volume of the material for transport. During this maturation, the ammonia will be oxidized into nitrates, improving the fertility of the material and making it more suitable as a soil improver. Large composting stages are typically used by dry anaerobic digestion technologies.

Wastewater

The final output from anaerobic digestion systems is water, which originates both from the moisture content of the original waste that was treated and water produced during the microbial reactions in the digestion systems. This water may be released from the dewatering of the digestate or may be implicitly separate from the digestate.

The wastewater exiting the anaerobic digestion facility will typically have elevated levels of biochemical oxygen demand (BOD) and chemical oxygen demand (COD). These measures of the reactivity of the effluent indicate an ability to pollute. Some of this material is termed 'hard COD', meaning it cannot be accessed by the anaerobic bacteria for conversion into biogas. If this effluent were put directly into watercourses, it would negatively affect them by causing eutrophication. As such, further treatment of the wastewater is often required. This treatment will typically be an oxidation stage wherein air is passed through the water in a sequencing batch reactors or reverse osmosis unit.

History

Gas street lamp

Reported scientific interest in the manufacturing of gas produced by the natural decomposition of organic matter dates from the 17th century, when Robert Boyle (1627-1691) and Stephen Hales (1677-1761) noted that disturbing the sediment of streams and lakes released flammable gas. In 1808 Sir Humphry Davy proved the presence of methane in the gases produced by cattle manure. In 1859 a leper colony in Bombay in India built the first anaerobic digester. In 1895, the technology was developed in Exeter, England, where a septic tank was used to generate gas for the sewer gas destructor lamp, a type of gas lighting. Also in England, in 1904, the first dual-purpose tank for both sedimentation and sludge treatment was installed in Hampton, London. In 1907, in Germany, a patent was issued for the Imhoff tank, an early form of digester.

Research on anaerobic digestion began in earnest in the 1930s.

Compost

Compost is organic matter that has been decomposed and recycled as a fertilizer and soil amendment. Compost is a key ingredient in organic farming.

A community-level composting plant in a rural area in Germany

At the simplest level, the process of composting simply requires making a heap of wetted organic matter known as green waste (leaves, food waste) and waiting for the materials to break down into humus after a period of weeks or months. Modern, methodical composting is a multi-step, closely monitored process with measured inputs of water, air, and carbon- and nitrogen-rich materials. The decomposition process is aided by shredding the plant matter, adding water and ensuring proper aeration by regularly turning the mixture. Worms and fungi further break up the material. Bacteria requiring oxygen to function (aerobic bacteria) and fungi manage the chemical process by converting the inputs into heat, carbon dioxide and ammonium. The ammonium (NH_4) is the form of nitrogen used by plants. When available ammonium is not used by plants it is further converted by bacteria into nitrates (NO_3) through the process of nitrification.

Compost is rich in nutrients. It is used in gardens, landscaping, horticulture, and agriculture. The compost itself is beneficial for the land in many ways, including as a soil conditioner, a fertilizer, addition of vital humus or humic acids, and as a natural pesticide for soil. In ecosystems, compost is useful for erosion control, land and stream reclamation, wetland construction, and as landfill cover. Organic ingredients intended for composting can alternatively be used to generate biogas through anaerobic digestion.

Terminology

Composting of waste is an aerobic (in the presence of air) method of decomposing solid wastes. The process involves decomposition of organic waste into humus known as compost which is a good fertiliser for plants. However, the term "composting" is used worldwide with differing meanings. Some composting textbooks narrowly define composting as being an aerobic form of decomposition, primarily by microbes. An alternative term to composting is "aerobic digestion", which in turn is also referred to as "wet composting".

For many people, composting is used to refer to several different types of biological process. In North America, "anaerobic composting" is still a common term for what much of the rest of the world and in technical publications people call "anaerobic digestion". The microbes used and the processes involved are quite different between composting and anaerobic digestion.

Ingredients

Home compost barrel in the Escuela Barreales, Santa Cruz, Chile

Carbon, Nitrogen, Oxygen, Water

Materials in a compost pile

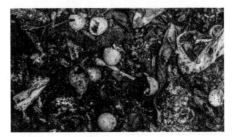

Food scraps compost heap

Composting organisms require four equally important ingredients to work effectively:

- Carbon — for energy; the microbial oxidation of carbon produces the heat, if included at suggested levels.

  o High carbon materials tend to be brown and dry.

- Nitrogen — to grow and reproduce more organisms to oxidize the carbon.

  o High nitrogen materials tend to be green (or colorful, such as fruits and vegetables) and wet.

- Oxygen — for oxidizing the carbon, the decomposition process.

- Water — in the right amounts to maintain activity without causing anaerobic conditions.

Certain ratios of these materials will provide beneficial bacteria with the nutrients to work at a rate that will heat up the pile. In that process much water will be released as vapor ("steam"), and the oxygen will be quickly depleted, explaining the need to actively manage the pile. The hotter the pile gets, the more often added air and water is necessary; the air/water balance is critical to maintaining high temperatures (135°-160° Fahrenheit / 50° - 70° Celsius) until the materials are broken down. At the same time, too much air or water also slows the process, as does too much carbon (or too little nitrogen). Hot container composting focuses on retaining the heat to increase decomposition rate and produce compost quicker.

The most efficient composting occurs with an optimal carbon:nitrogen ratio of about 10:1 to 20:1. Rapid composting is favored by having a C/N ratio of ~30 or less. Theoretical analysis is confirmed by field tests that above 30 the substrate is nitrogen starved, below 15 it is likely to outgas a portion of nitrogen as ammonia. If nitrogen needs to be increased, it has been suggested to add 0.15 pounds of *actual* nitrogen per three bushels (3.75 cubic feet) of lower nitrogen material. [For those not familiar with these types of units: 0.64g/L or 640 grams of actual nitrogen per cubic meter.] Two to 3 pounds of organic nitrogen supplement (blood meal, manure, bone meal, alfalfa meal) per 100 pounds of low nitrogen materials (for example, straw or sawdust), supplies generally ample nitrogen and trace minerals in high carbon mixes.

Nearly all plant and animal materials have both carbon and nitrogen, but amounts vary widely, with characteristics noted above (dry/wet, brown/green). Fresh grass clippings have an average ratio of about 15:1 and dry autumn leaves about 50:1 depending on species. Mixing equal parts by volume approximates the ideal C:N range. Few individual situations will provide the ideal mix of materials at any point. Observation of amounts, and consideration of different materials as a pile is built over time, can quickly achieve a workable technique for the individual situation.

Animal Manure and Bedding

On many farms, the basic composting ingredients are animal manure generated on the farm and bedding. Straw and sawdust are common bedding materials. Non-traditional bedding materials are also used, including newspaper and chopped cardboard. The amount of manure composted on a livestock farm is often determined by cleaning schedules, land availability, and weather conditions. Each type of manure has its own physical, chemical, and biological characteristics. Cattle and horse manures, when mixed with bedding, possess good qualities for composting. Swine manure, which is very wet and usually not mixed with bedding material, must be mixed with straw or similar raw materials. Poultry manure also must be blended with carbonaceous materials - those low in nitrogen preferred, such as sawdust or straw.

Microorganisms

With the proper mixture of water, oxygen, carbon, and nitrogen, micro-organisms are allowed to break down organic matter to produce compost. The composting process is dependent on micro-organisms to break down organic matter into compost. There are many types of microorganisms found in active compost of which the most common are:

- Bacteria- The most numerous of all the microorganisms found in compost. Depending on the phase of composting, mesophilic or thermophilic bacteria may predominate.

- Actinobacteria- Necessary for breaking down paper products such as newspaper, bark, etc.

- Fungi- Molds and yeast help break down materials that bacteria cannot, especially lignin in woody material.

- Protozoa- Help consume bacteria, fungi and micro organic particulates.

- Rotifers- Rotifers help control populations of bacteria and small protozoans.

In addition, earthworms not only ingest partly composted material, but also continually re-create aeration and drainage tunnels as they move through the compost.

A lack of a healthy micro-organism community is the main reason why composting processes are slow in landfills with environmental factors such as lack of oxygen, nutrients or water being the cause of the depleted biological community.

Phases of Composting

Under ideal conditions, composting proceeds through three major phases:

- An initial, mesophilic phase, in which the decomposition is carried out under moderate temperatures by mesophilic microorganisms.

- As the temperature rises, a second, thermophilic phase starts, in which the decomposition is carried out by various thermophilic bacteria under high temperatures.

- As the supply of high-energy compounds dwindles, the temperature starts to decrease, and the mesophiles once again predominate in the maturation phase.

Human Waste

Human waste (excreta) can also be added as an input to the composting process, like it is done in composting toilets, as human waste is a nitrogen-rich organic material.

People excrete far more water-soluble plant nutrients (nitrogen, phosphorus, potassium) in urine than in feces. Human urine can be used directly as fertilizer or it can be put onto compost. Adding a healthy person's urine to compost usually will increase temperatures and therefore increase its ability to destroy pathogens and unwanted seeds. Urine from a person with no obvious symptoms of infection is much more sanitary than fresh feces. Unlike feces, urine does not attract disease-spreading flies (such as house flies or blow flies), and it does not contain the most hardy of pathogens, such as parasitic worm eggs. Urine usually does not stink for long, particularly when it is fresh, diluted, or put on sorbents.

Urine is primarily composed of water and urea. Although metabolites of urea are nitrogen fertilizers, it is easy to over-fertilize with urine, or to utilize urine containing pharmaceutical (or other) content, creating too much ammonia for plants to absorb, acidic conditions, or other phytotoxicity.

Humanure

"Humanure" is a portmanteau of *human* and *manure*, designating human excrement (feces and urine) that is recycled via composting for agricultural or other purposes. The term was first used in a 1994 book by Joseph Jenkins that advocates the use of this organic soil amendment. The term humanure is used by compost enthusiasts in the US but not generally elsewhere. Because the term "humanure" has no authoritative definition it is subject to various uses; news reporters occasionally fail to correctly distinguish between humanure and sewage sludge or "biosolids".

Uses

Compost is generally recommended as an additive to soil, or other matrices such as coir and peat, as a tilth improver, supplying humus and nutrients. It provides a rich *growing medium*, or a porous, absorbent material that holds moisture and soluble minerals, providing the support and nutrients in which plants can flourish, although it is rarely used alone, being primarily mixed with soil, sand, grit, bark chips, vermiculite, perlite, or clay granules to produce loam. Compost can be tilled directly into the soil or growing medium to boost the level of organic matter and the overall fertility of the soil. Compost that is ready to be used as an additive is dark brown or even black with an earthy smell.

Generally, direct seeding into a compost is not recommended due to the speed with which it may dry and the possible presence of phytotoxins that may inhibit germination, and the possible tie up of nitrogen by incompletely decomposed lignin. It is very common to see blends of 20–30% compost used for transplanting seedlings at cotyledon stage or later.

Composting can destroy pathogens or unwanted seeds. Unwanted living plants (or weeds) can be discouraged by covering with mulch/compost. The "microbial pesticides" in compost may include thermophiles and mesophiles, however certain composting detritivores such as black soldier fly larvae and redworms, also reduce many pathogens. Thermophilic (high-temperature) composting is well known to destroy many seeds and nearly all types of pathogens (exceptions may include prions). The sanitizing qualities of (thermophilic) composting are desirable where there is a high likelihood of pathogens, such as with manure.

Composting Technologies

A modern compost bin constructed from plastics   A homemade compost tumbler

Overview

In addition to the traditional compost pile, various approaches have been developed to handle different composting processes, ingredients, locations, and applications for the composted product.

There is a large number of different composting systems on the market, for example:

- At the household level: Composting toilet, container composting, vermicomposting
- At the industrial composting (large scale): Aerated Static Pile Composting, vermicomposting, windrow composting etc.

Examples

Vermicomposting

Vermicompost is the product or process of composting through the utilization of various species of worms, usually red wigglers, white worms, and earthworms, to create a heterogeneous mixture of decomposing vegetable or food waste (excluding meat, dairy, fats, or oils), bedding materials, and vermicast. Vermicast, also known as worm castings, worm humus or worm manure, is the end-product of the breakdown of organic matter by species of earthworm. Vermicomposting is widely used in North America for on-site institutional processing of food waste, such as in hospitals and shopping malls. This type of composting is sometimes suggested as a feasible indoor home composting method. Vermicomposting has gained popularity in both these industrial and domestic settings because, as compared with conventional composting, it provides a way to compost organic materials more quickly (as defined by a higher rate of carbon-to-nitrogen ratio increase) and to attain products that have lower salinity levels that are therefore more beneficial to plant mediums.

The earthworm species (or composting worms) most often used are red wigglers (*Eisenia fetida* or *Eisenia andrei*), though European nightcrawlers (*Eisenia hortensis* or *Dendrobaena veneta*) could also be used. Red wigglers are recommended by most vermiculture experts, as they have some of the best appetites and breed very quickly. Users refer to European nightcrawlers by a variety of other names, including *dendrobaenas*, *dendras*, Dutch Nightcrawlers, and Belgian nightcrawlers.

Containing water-soluble nutrients, vermicompost is a nutrient-rich organic fertilizer and soil conditioner in a form that is relatively easy for plants to absorb. Worm castings are sometimes used as an organic fertilizer. Because the earthworms grind and uniformly mix minerals in simple forms, plants need only minimal effort to obtain them. The worms' digestive systems also add beneficial microbes to help create a "living" soil environment for plants.

Rotary screen harvested worm castings

Vermicompost tea in conjunction with 10% castings has been shown to cause up to a 1.7 times growth in plant mass over plants grown without.

Food waste - after three years

Researchers from the Pondicherry University discovered that worm composts can also be used to clean up heavy metals. The researchers found substantial reductions in heavy metals when the worms were released into the garbage and they are effective at removing lead, zinc, cadmium, copper and manganese.

Hügelkultur (Raised Garden Beds or Mounds)

The practice of making raised garden beds or mounds filled with rotting wood is also called "Hügelkultur" in German. It is in effect creating a Nurse log that is covered with dirt.

Benefits of hügelkultur garden beds include water retention and warming of soil. Buried wood becomes like a sponge as it decomposes, able to capture water and store it for later use by crops planted on top of the hügelkultur bed.

The buried decomposing wood will also give off heat, as all compost does, for several years. These effects have been used by Sepp Holzer to enable fruit trees to survive at otherwise inhospitable temperatures and altitudes.

An almost completed Hügelkultur bed; the bed does not have dirt on it yet.

Black Soldier Fly Larvae Composting

Black Soldier Fly (*Hermetia illucens*) larvae have been shown to be able to rapidly consume large amounts of organic waste when kept at 31.8 °C, the optimum temperature for reproduction. Enthusiasts have experimented with a large number of different waste products and some even sell starter kits to the public.

Cockroach Composting

Cockroach composting is another insect-mediated composting method. In this case the adults of any number of cockroach species (such as the Turkestan cockroach or *Blaptica dubia*) are used to quickly convert manure or kitchen waste to nutrient dense compost. Depending on species used and environmental conditions, excess composting insects can be used as an excellent animal feed for farm animals and pets.

Bokashi

Bokashi is a method that uses a mix of microorganisms to cover food waste or wilted plants to decrease smell. Bokashi (ぼかし) is Japanese for "shading off" or "gradation." It derives from the practice of Japanese farmers centuries ago of covering food waste with rich, local soil that contained the microorganisms that would ferment the waste. After a few weeks, they would bury the waste.

Inside a recently started bokashi bin. The aerated base is just visible through the food scraps and bokashi bran.

Most practitioners obtain the microorganisms from the product Effective Microorganisms (EM1), first sold in the 1980s. EM1 is mixed with a carbon base (e.g. sawdust or bran) that it sticks to and a sugar for food (e.g. molasses). The mixture is layered with waste in a sealed container and after a few weeks, removed and buried.

Newspaper fermented in a lactobacillus culture can be substituted for bokashi bran for a successful bokashi bucket.

Compost Tea

Compost teas are defined as water extracts brewed from composted materials and can be derived from aerobic or anaerobic processes. Compost teas are generally produced from adding one volume of compost to 4-10 volumes of water, but there has also been debate about the benefits of aerating the mixture. Field studies have shown the benefits of adding compost teas to crops due to the adding of organic matter, increased nutrient availability and increased microbial activity. They have also been shown to have an effect on plant pathogens.

Composting Toilets

A composting toilet does not require water or electricity, and when properly managed does not smell. A composting toilet collects human excreta which is then added to a compost heap together with sawdust and straw or other carbon rich materials, where pathogens are destroyed to some extent. The amount of pathogen destruction depends on the temperature (mesophilic or thermophilic conditions) and composting time. A composting toilet tries to process the excreta in situ although this is often coupled with a secondary external composting step. The resulting compost product has been given various names, such as humanure and EcoHumus.

A composting toilet can aid in the conservation of fresh water by avoiding the usage of potable water required by the typical flush toilet. It further prevents the pollution of ground water by controlling the fecal matter decomposition before entering the system. When properly managed, there should be no ground contamination from leachate.

Compost and Land-Filling

As concern about landfill space increases, worldwide interest in recycling by means of composting is growing, since composting is a process for converting decomposable organic materials into useful stable products. Composting is one of the only ways to revitalize soil vitality due to phosphorus depletion in soil. Industrial scale composting in the form of in-vessel composting, aerated static pile composting, and anaerobic digestion takes place in most Western countries now, and in many areas is mandated by law. There are process and product guidelines in Europe that date to the early 1980s (Germany, the Netherlands, Switzerland) and only more recently in the UK and the US. In both these countries, private trade associations within the industry have established loose standards, some say as a stop-gap measure to discourage independent government agencies from establishing tougher consumer-friendly standards. The USA is the only Western country that does not distinguish sludge-source compost from green-composts, and by default in the USA 50% of states expect composts to comply in some manner with the federal EPA 503 rule promulgated in 1984 for sludge products. Compost is regulated in Canada and Australia as well.

Industrial Systems

Industrial composting systems are increasingly being installed as a waste management alternative to landfills, along with other advanced waste processing systems. Mechanical sorting of mixed waste streams combined with anaerobic digestion or in-vessel composting is called mechanical biological treatment, and is increasingly being used in developed countries due to regulations controlling the amount of organic matter allowed in landfills. Treating biodegradable waste before it enters a landfill reduces global warming from fugitive methane; untreated waste breaks down anaerobically in a landfill, producing landfill gas that contains methane, a potent greenhouse gas.

A large compost pile that is steaming with the heat generated by thermophilic microorganisms.

Vermicomposting, also known as vermiculture, is used for medium-scale on-site institutional composting, such as for food waste from universities and shopping malls. It is selected either as a more environmentally friendly choice than conventional methods of disposal, or to reduce the cost of commercial waste removal.

Large-scale composting systems are used by many urban areas around the world. Co-composting is a technique that combines solid waste with de-watered biosolids, although difficulties controlling inert and plastics contamination from municipal solid waste makes this approach less attractive. The world's largest MSW co-composter is the Edmonton Composting Facility in Edmonton, Alberta, Canada, which turns 220,000 tonnes of residential solid waste and 22,500 dry tonnes of biosolids per year into 80,000 tonnes of compost. The facility is 38,690 m² (416,500 sq.ft.) in area, equivalent to 4½ Canadian football fields, and the operating structure is the largest stainless steel building in North America, the size of 14 NHL rinks. In 2006, Qatar awarded Keppel Seghers Singapore, a subsidiary of Keppel Corporation, a contract to begin construction on a 275,000 tonne/year anaerobic digestion and composting plant licensed by Kompogas (de) Switzerland. This plant, with 15 independent anaerobic digesters, will be the world's largest composting facility once fully operational in early 2011 and forms part of Qatar's Domestic Solid Waste Management Centre, the largest integrated waste management complex in the Middle East.

Another large MSW composter is the Lahore Composting Facility in Lahore, Pakistan, which has a capacity to convert 1,000 tonnes of municipal solid waste per day into compost. It also has a capacity to convert substantial portion of the intake into refuse-derived fuel (RDF) materials for further combustion use in several energy consuming industries across Pakistan, for example in cement manufacturing companies where it is used to heat cement kilns. This project has also been approved by the Executive Board of the United Nations Framework Convention on Climate Change for reducing methane emissions, and has been registered with a capacity of reducing 108,686 tonnes CO_2 equivalent per annum.

Related Technologies

Anaerobic digestion is process for converting organic waste into (biogas). The residual material, sometimes in combination with sewage sludge can be followed by an aerobic composting process before selling or giving away the compost.

History

Composting as a recognized practice dates to at least the early Roman Empire since Pliny the Elder (AD 23-79). Traditionally, composting involved piling organic materials until the next planting season, at which time the materials would have decayed enough to be ready for use in the soil. The advantage of this method is that little working time or effort is required from the composter and it fits in naturally with agricultural practices in temperate climates. Disadvantages (from the modern perspective) are that space is used for a whole year, some nutrients might be leached due to exposure to rainfall, and disease-producing organisms and insects may not be adequately controlled.

Compost Basket

Composting was somewhat modernized beginning in the 1920s in Europe as a tool for organic farming. The first industrial station for the transformation of urban organic materials into compost was set up in Wels, Austria in the year 1921. Early frequent citations for propounding composting within farming are for the German-speaking world Rudolf Steiner, founder of a farming method called biodynamics, and Annie Francé-Harrar, who was appointed on behalf of the government in Mexico and supported the country 1950–1958 to set up a large humus organization in the fight against erosion and soil degradation.

In the English-speaking world it was Sir Albert Howard who worked extensively in India on sustainable practices and Lady Eve Balfour who was a huge proponent of composting. Composting was imported to America by various followers of these early European movements by the likes of J.I. Rodale (founder of Rodale Organic Gardening), E.E. Pfeiffer (who developed scientific practices in biodynamic farming), Paul Keene (founder of Walnut Acres in Pennsylvania), and Scott and Helen Nearing (who inspired the back-to-the-land movement of the 1960s). Coincidentally, some of the above met briefly in India - all were quite influential in the U.S. from the 1960s into the 1980s.

There are many modern proponents of rapid composting that attempt to correct some of the perceived problems associated with traditional, slow composting. Many advocate that compost can be made in 2 to 3 weeks. Many such short processes involve a few changes to traditional methods, including smaller, more homogenized pieces in the compost, controlling carbon-to-nitrogen ratio (C:N) at 30 to 1 or less, and monitoring the moisture level more carefully. However, none of these parameters differ significantly from the early writings of Howard and Balfour, suggesting that in fact modern composting has not made significant advances over the traditional methods that take a few months to work. For this reason and others, many modern scientists who deal with carbon transformations are sceptical that there is a "super-charged" way to get nature to make compost rapidly.

In fact, both sides are right to some extent. The bacterial activity in rapid high heat methods breaks down the material to the extent that pathogens and seeds are destroyed, and the original feedstock is unrecognizable. At this stage, the compost can be used to prepare fields or other planting areas. However, most professionals recommend that the compost be given time to cure before using in a nursery for starting seeds or growing young plants. The curing time allows fungi to continue the decomposition process and eliminating phytotoxic substances.

Many countries such as Wales and some individual cities such as Seattle and San Francisco require food and yard waste to be sorted for composting.

Kew Gardens in London has one of the biggest non-commercial compost heaps in Europe.

Photodegradation

Photo-degradation is the alteration of materials by light. Typically, the term refers to the combined action of sunlight and air. Photo-degradation is usually oxidation and hydrolysis. Often photodegradation is avoided, since it destroys paintings and other artifacts. It is however partly responsible for remineralization of biomass and is used intentionally in some disinfection technologies. Photodegridation does not apply to how materials may be aged or degraded via Infrared light or heat, but does include degredation in all of the ultravioliet light wavebands.

Where Photodegradation is Important

Foodstuffs

The protection of food from photodegradation is very important. Some nutrients, like beer for example, are affected by degradation when being exposed to sunlight. In the case of beer, the UV radiation causes a process, which entails the degradation of hop bitter compounds to 3-Methyl-2-buten-1-thiol and therefore changes the taste. As amber glass has the ability to absorb UV radiation, beer bottles are often made from this glass type to avoid this process.

Paints, Inks And Dyes

Paints, inks and dyes that are organic are more subseptable to photodegredation than those that are not. Ceramics are almost universally coloured with non-organic origin materials so as to allow the material to retain its colour even with the most relentless photo degredation.

Pesticides & Herbicides

The photodegradation of pesticides is of great interest because of the scale of agriculture and the intensive use of chemicals. Pesticides are however selected in part not to photodegrade readily in sunlight in order to allow them to exert their biocidal activity. Thus, additional modalities are implemented to enhance their photodegradation, including the use of photosensitizers, photocatalysts (e.g., titanium dioxide), and the addition of reagents such as hydrogen peroxide that would generate hydroxyl radicals that would attack the pesticides.

Pharmaceuticals

The photodegradation of pharmaceuticals is of interest because they are found in many water supplies. They have deleterious effects on aquatic organisms including toxicity, endocrine disruption, genetic damage. But also in the primary packaging material the photodegradation of pharmaceuticals has to be prevented. For this, amber glasses like FIOLAX® amber are commonly used to protect the pharmaceutical from UV radiations. Iodine (in the form of Lugol's Solution) and collodial silver are universally used in packaging that lets through very little UV light so as to avoid degradation.

Polymers

Common synthetic polymers that can be attacked include polypropylene and LDPE, where tertiary carbon bonds in their chain structures are the centres of attack. Ultraviolet rays interact with these bonds to form free radicals, which then react further with oxygen in the atmosphere, producing carbonyl groups in the main chain. The exposed surfaces of products may then discolour and crack, and in extreme cases, complete product disintegration can occur.

Effect of UV exposure on polypropylene rope

In fibre products like rope used in outdoor applications, product life will be low because the outer fibres will be attacked first, and will easily be damaged by abrasion for example. Discolouration of the rope may also occur, thus giving an early warning of the problem.

Polymers which possess UV-absorbing groups such as aromatic rings may also be sensitive to UV degradation. Aramid fibres like Kevlar, for example, are highly UV-sensitive and must be protected from the deleterious effects of sunlight.

Mechanism of Photodegradation

Many organic chemicals are thermodynamically unstable in the presence of oxygen, however, their rate of spontaneous oxidation is slow at room temperature. In the language of physical chemistry, such reactions are kinetically limited. This kinetic stability allows the accumulation of complex environmental structures in the environment. Upon the absorption of light, triplet oxygen converts to singlet oxygen, a highly reactive form of the gas, which effects spin-allowed oxidations. In the atmosphere, the organic compounds are degraded by hydroxyl radicals, which are produced from water and ozone.

Photodegradation of a plastic bucket used as an open-air flowerpot for some years

Photochemical reactions are initiated by the absorption of a photon, typically in the wavelength range 290-700 nm (at the surface of the Earth). The energy of an absorbed photon is transferred to electrons in the molecule and briefly changes their configuration (i.e., promotes the molecule from a ground state to an excited state). The excited state represents what is essentially a new molecule. Often excited state molecules are not kinetically stable in the presence of O_2 or H_2O and can spontaneously decompose (oxidize or hydrolyze). Sometimes molecules decompose to produce high energy, unstable fragments that can react with other molecules around them. The two processes are collectively referred to as direct photolysis or indirect photolysis, and both mechanisms contribute to the removal of pollutants.

The United States federal standard for testing plastic for photo-degradation is 40 CFR Ch. I (7–1–03 Edition)PART 238

Protection Against Photodegradation

Photodegradation of plastics and other materials can be inhibited with additives, which are widely used. These additives include antioxidants, which interrupt degradation processes. Typical antioxidants are derivatives of aniline. Another type of additive are UV-absorbers. These agents capture the photon and convert it to heat. Typical UV-absorbers are hydroxy-substituted benzophenones, related to the chemicals used in sunscreen.

Bioconversion

Bioconversion, also known as *biotransformation*, is the conversion of organic materials, such as plant or animal waste, into usable products or energy sources by biological processes or agents, such as certain microorganisms. One example is the industrial production of cortisone, which one step is the bioconversion of progesterone to 11-alpha-Hydroxyprogesterone by *Rhizopus nigricans*. Another example is the bioconversion of glycerol to 1,3-propanediol, which is part of scientific research for many decades.

Another example of bioconversion is the conversion of organic materials, such as plant or animal waste, into usable products or energy sources by biological processes or agents, such as certain microorganisms, some detritivores or enzymes.

In the USA, the Bioconversion Science and Technology group performs multidisciplinary R&D for the Department of Energy's (DOE) relevant applications of bioprocessing, especially with biomass. Bioprocessing combines the disciplines of chemical engineering, microbiology and biochemistry. The Group 's primary role is investigation of the use of microorganism, microbial consortia and microbial enzymes in bioenergy research.

New cellulosic ethanol conversion processes have enabled the variety and volume of feedstock that can be bioconverted to expand rapidly. Feedstock now includes materials derived from plant or animal waste such as paper, auto-fluff, tires, fabric, construction materials, municipal solid waste (MSW), sludge, sewage, etc.

Three Different Processes for Bioconversion

1 - Enzymatic hydrolysis - a single source of feedstock, switchgrass for example, is mixed with strong enzymes which convert a portion of cellulosic material into sugars which can then be fermented into ethanol. Genencor and Novozymes are two companies that have received United States government Department of Energy funding for research into reducing the cost of cellulase, a key enzyme in the production cellulosic ethanol by this process.

2 - Synthesis gas fermentation - a blend of feedstock, not exceeding 30% water, is gasified in a closed environment into a syngas containing mostly carbon monoxide and hydrogen. The cooled syngas is then converted into usable products through exposure to bacteria or other catalysts. BRI Energy, LLC is a company whose pilot plant in Fayetteville, Arkansas is currently using synthesis gas fermentation to convert a variety of waste into ethanol. After gasification, anaerobic bacteria (*Clostridium ljungdahlii*) are used to convert the syngas (CO, CO_2, and H_2) into ethanol. The heat generated by gasification is also used to co-generate excess electricity.

3 - C.O.R.S. and Grub Composting are sustainable technologies that employ organisms that feed on organic matter to reduce and convert organic waste in to high quality feedstuff and oil rich material for the biodiesel industry. Organizations pioneering this novel approach to waste man-

agement are EAWAG, ESR International, Prota Culture and BIOCONVERSION that created the *e*-CORS® system to meet large scale organic waste management needs and environmental sustainability in both urban and livestock farming reality. This type of engineered system introduces a substantial innovation represented by the automatic modulation of the treatment, able to adapt conditions of the system to the biology of the scavenger used, improving their performances and the power of this technology.

Decomposition

Decomposition is the process by which organic substances are broken down into much simpler forms of matter. The process is essential for recycling the finite matter that occupies physical space in the biome. Bodies of living organisms begin to decompose shortly after death. Animals, such as worms, also help decompose the organic materials. Organisms that do this are known as decomposers. Although no two organisms decompose in the same way, they all undergo the same sequential stages of decomposition. The science which studies decomposition is generally referred to as *taphonomy*.

One can differentiate abiotic from biotic decomposition (biodegradation). The former means "degradation of a substance by chemical or physical processes, e.g., hydrolysis. The latter means "the metabolic breakdown of materials into simpler components by living organisms", typically by microorganisms.

Animal Decomposition

Decomposition begins at the moment of death, caused by two factors: autolysis, the breaking down of tissues by the body's own internal chemicals and enzymes, and putrefaction, the breakdown of tissues by bacteria. These processes release gases, such as cadaverine and putrescine, that are the chief source of the unmistakably putrid odor of decaying animal tissue.

Ants eating a dead snake

Prime decomposers are bacteria or fungi, though larger scavengers also play an important role in decomposition if the body is accessible to insects, mites and other animals. The most important arthropods that are involved in the process include carrion beetles, mites, the flesh-flies (Sarcophagidae) and blow-flies (Calliphoridae), such as the green-bottle fly seen in the summer. The most important non-insect animals that are typically involved in the process include mammal and bird scavengers, such as coyotes, dogs, wolves, foxes, rats, crows and vultures. Some of these scavengers also remove and scatter bones, which they ingest at a later time. Aquatic and marine environments have break-down agents that include bacteria, fish, crustaceans, Diptera larvae and other carrion scavengers.

Stages of Decomposition

Five general stages are used to describe the process of decomposition in vertebrate animals: fresh, bloat, active and advanced decay, and dry/remains. The general stages of decomposition are coupled with two stages of chemical decomposition: autolysis and putrefaction. These two stages contribute to the chemical process of decomposition, which breaks down the main components of the body.

Fresh

The fresh stage begins immediately after the heart stops beating. A person can be revived, especially if someone with Cardiopulmonary resuscitation skills is nearby. From the moment of death, the body begins cooling or warming to match the temperature of the ambient environment, during a stage called algor mortis. Shortly after death, within three to six hours, the muscular tissues become rigid and incapable of relaxing, during a stage called rigor mortis. Since blood is no longer being pumped through the body, gravity causes it to drain to the dependent portions of the body, creating an overall bluish-purple discolouration termed livor mortis or, more commonly, lividity.

Once the heart stops, the blood can no longer supply oxygen or remove carbon dioxide from the tissues. The resulting decrease in pH and other chemical changes cause cells to lose their structural integrity, bringing about the release of cellular enzymes capable of initiating the breakdown of surrounding cells and tissues. This process is known as autolysis.

Visible changes caused by decomposition are limited during the fresh stage, although autolysis may cause blisters to appear at the surface of the skin.

The small amount of oxygen remaining in the body is quickly depleted by cellular metabolism and aerobic microbes naturally present in respiratory and gastrointestinal tracts, creating an ideal environment for the proliferation of anaerobic organisms. These multiply, consuming the body's carbohydrates, lipids, and proteins, to produce a variety of substances including propionic acid, lactic acid, methane, hydrogen sulfide, and ammonia. The process of microbial proliferation within a body is referred to as putrefaction and leads to the second stage of decomposition, known as bloat.

Blowflies and flesh flies are the first carrion insects to arrive, and they seek a suitable oviposition site.

Bloat

The bloat stage provides the first clear visual sign that microbial proliferation is underway. In this stage, anaerobic metabolism takes place, leading to the accumulation of gases, such as hydrogen sulfide, carbon dioxide, methane, and nitrogen. The accumulation of gases within the bodily cavity causes the distention of the abdomen and gives a cadaver its overall bloated appearance. The gases produced also cause natural liquids and liquefying tissues to become frothy. As the pressure of the gases within the body increases, fluids are forced to escape from natural orifices, such as the nose, mouth, and anus, and enter the surrounding environment. The buildup of pressure combined with the loss of integrity of the skin may also cause the body to rupture.

Intestinal anaerobic bacteria transform haemoglobin into sulfhemoglobin and other colored pigments. The associated gases which accumulate within the body at this time aid in the transport of sulfhemoglobin throughout the body via the circulatory and lymphatic systems, giving the body an overall marbled appearance.

If insects have access, maggots hatch and begin to feed on the body's tissues. Maggot activity, typically confined to natural orifices, and masses under the skin, causes the skin to slip, and hair to detach from the skin. Maggot feeding, and the accumulation of gases within the body, eventually leads to post-mortem skin ruptures which will then further allow purging of gases and fluids into the surrounding environment. Ruptures in the skin allow oxygen to re-enter the body and provide more surface area for the development of fly larvae and the activity of aerobic microorganisms. The purging of gases and fluids results in the strong distinctive odors associated with decay.

Active Decay

Active decay is characterized by the period of greatest mass loss. This loss occurs as a result of both the voracious feeding of maggots and the purging of decomposition fluids into the surrounding environment. The purged fluids accumulate around the body and create a cadaver decomposition island (CDI). Liquefaction of tissues and disintegration become apparent during this time and strong odors persist. The end of active decay is signaled by the migration of maggots away from the body to pupate.

Advanced Decay

Decomposition is largely inhibited during advanced decay due to the loss of readily available cadaveric material. Insect activity is also reduced during this stage. When the carcass is located on soil, the area surrounding it will show evidence of vegetation death. The CDI surrounding the carcass will display an increase in soil carbon and nutrients, such as phosphorus, potassium, calcium, and magnesium; changes in pH; and a significant increase in soil nitrogen.

Dry/Remains

During the dry/remains stage, the resurgence of plant growth around the CDI may occur and is a sign that the nutrients present in the surrounding soil have not yet returned to their normal levels. All that remains of the cadaver at this stage is dry skin, cartilage, and bones, which will become dry and bleached if exposed to the elements. If all soft tissue is removed from the cadaver, it is referred to as completely skeletonized, but if only portions of the bones are exposed, it is referred to as partially skeletonised.

Pig carcass in the different stages of decomposition: Fresh > Bloat > Active decay > Advanced decay > Dry remains

Plant Decomposition

Decomposition of plant matter occurs in many stages. It begins with leaching by water; the most easily lost and soluble carbon compounds are liberated in this process. Another early process is physical breakup or fragmentation of the plant material into smaller bits which have greater surface area for microbial colonization and attack. In smaller dead plants, this process is largely carried out by the soil invertebrate fauna, whereas in the larger plants, primarily parasitic life-forms such as insects and fungi play a major breakdown role and are not assisted by numerous detritivore species. Following this, the plant detritus (consisting of cellulose, hemicellulose, microbial products, and lignin) undergoes chemical alteration by microbes. Different types of compounds decompose at different rates. This is dependent on their chemical structure. For instance, lignin is a component of wood, which is relatively resistant to decomposition and can in fact only be decomposed by certain fungi, such as the black-rot fungi. Said fungi are thought to be seeking the nitrogen content of lignin rather than its carbon content. Lignin is one such remaining product of decomposing plants with a very complex chemical structure causing the rate of microbial breakdown to slow. Warmth determines the speed of plant decay, with the rate of decay increasing as heat increases, i.e. a plant in a warm environment will decay over a shorter period of time. In most grassland ecosystems, natural damage from fire, insects that feed on decaying matter, termites, grazing mammals, and the physical movement of animals through the grass are the primary agents of breakdown and nutrient cycling, while bacteria and fungi play the main roles in further decomposition.

A decaying peach over a period of six days. Each frame is approximately 12 hours apart, as the fruit shrivels and becomes covered with mold.

The chemical aspects of plant decomposition always involve the release of carbon dioxide.

Food Decomposition

The decomposition of food, either plant or animal, called *spoilage* in this context, is an important field of study within food science. Food decomposition can be slowed down by conservation. The spoilage of meat occurs, if the meat is untreated, in a matter of hours or days and results in the meat becoming unappetizing, poisonous or infectious. Spoilage is caused by the practically unavoidable infection and subsequent decomposition of meat by bacteria and fungi, which are borne by the animal itself, by the people handling the meat, and by their implements. Meat can be kept edible for a much longer time – though not indefinitely – if proper hygiene is observed during production and processing, and if appropriate food safety, food preservation and food storage procedures are applied.

Importance to Forensics

Various sciences study the decomposition of bodies under the general rubric of forensics because the usual motive for such studies is to determine the time and cause of death for legal purposes:

- Forensic taphonomy specifically studies the processes of decomposition in order to apply the biological and chemical principles to forensic cases in order to determine post-mortem interval (PMI), post-burial interval as well as to locate clandestine graves.

- Forensic pathology studies the clues to the cause of death found in the corpse as a medical phenomenon.

- Forensic entomology studies the insects and other vermin found in corpses; the sequence in which they appear, the kinds of insects, and where they are found in their life cycle are clues that can shed light on the time of death, the length of a corpse's exposure, and whether the corpse was moved.

- Forensic anthropology is the branch of physical anthropology that studies skeletons and human remains, usually to seek clues as to the identity, race, and sex of their former owner.

The University of Tennessee Anthropological Research Facility (better known as the Body Farm) in Knoxville, Tennessee has a number of bodies laid out in various situations in a fenced-in plot near the medical center. Scientists at the Body Farm study how the human body decays in various circumstances to gain a better understanding of decomposition.

Factors Affecting Decomposition

Exposure to The Elements

A dead body that has been exposed to the open elements, such as water and air, will decompose more quickly and attract much more insect activity than a body that is buried or confined in special protective gear or artifacts. This is due, in part, to the limited number of insects that can penetrate a coffin and the lower temperatures under soil.

The rate and manner of decomposition in an animal body is strongly affected by several factors. In roughly descending degrees of importance, they are:

- Temperature;
- The availability of oxygen;
- Prior embalming;
- Cause of death;
- Burial, depth of burial, and soil type;
- Access by scavengers;
- Trauma, including wounds and crushing blows;
- Humidity, or wetness;

- Rainfall;

- Body size and weight;

- Clothing;

- The surface on which the body rests;

- Foods/objects inside the specimen's digestive tract (bacon compared to lettuce).

The speed at which decomposition occurs varies greatly. Factors such as temperature, humidity, and the season of death all determine how fast a fresh body will skeletonize or mummify. A basic guide for the effect of environment on decomposition is given as Casper's Law (or Ratio): if all other factors are equal, then, when there is free access of air a body decomposes twice as fast than if immersed in water and eight times faster than if buried in earth. Ultimately, the rate of bacterial decomposition acting on the tissue will be depend upon the temperature of the surroundings. Colder temperatures decrease the rate of decomposition while warmer temperatures increase it.

The most important variable is a body's accessibility to insects, particularly flies. On the surface in tropical areas, invertebrates alone can easily reduce a fully fleshed corpse to clean bones in un-der two weeks. The skeleton itself is not permanent; acids in soils can reduce it to unrecognizable components. This is one reason given for the lack of human remains found in the wreckage of the *Titanic*, even in parts of the ship considered inaccessible to scavengers. Freshly skeletonized bone is often called "green" bone and has a characteristic greasy feel. Under certain conditions (nor-mally cool, damp soil), bodies may undergo saponification and develop a waxy substance called adipocere, caused by the action of soil chemicals on the body's proteins and fats. The formation of adipocere slows decomposition by inhibiting the bacteria that cause putrefaction.

In extremely dry or cold conditions, the normal process of decomposition is halted – by either lack of moisture or temperature controls on bacterial and enzymatic action – causing the body to be preserved as a mummy. Frozen mummies commonly restart the decomposition process when thawed, whilst heat-desiccated mummies remain so unless exposed to moisture.

The bodies of newborns who never ingested food are an important exception to the normal process of decomposition. They lack the internal microbial flora that produce much of decomposition and quite commonly mummify if kept in even moderately dry conditions.

Artificial Preservation

Embalming is the practice of delaying decomposition of human and animal remains. Embalming slows decomposition somewhat, but does not forestall it indefinitely. Embalmers typically pay great attention to parts of the body seen by mourners, such as the face and hands. The chemicals used in embalming repel most insects, and slow down bacterial putrefaction by either killing existing bacteria in or on the body themselves or by "fixing" cellular proteins, which means that they cannot act as a nutrient source for subsequent bacterial infections. In sufficiently dry environments, an embalmed body may end up mummified and it is not un-common for bodies to remain preserved to a viewable extent after decades. Notable viewable embalmed bodies include those of:

- Eva Perón of Argentina, whose body was injected with paraffin was kept perfectly preserved for many years, and still is as far as is known (her body is no longer on public display).

- Vladimir Lenin of the Soviet Union, whose body was kept submerged in a special tank of fluid for decades and is on public display in Lenin's Mausoleum.

    o Other Communist leaders with pronounced cults of personality such as Mao Zedong, Kim Il-sung, Ho Chi Minh, Kim Jong-il and most recently Hugo Chávez have also had their cadavers preserved in the fashion of Lenin's preservation and are now displayed in their respective mausoleums.

- Pope John XXIII, whose preserved body can be viewed in St. Peter's Basilica.

- Padre Pio, whose body was injected with formalin prior to burial in a dry vault from which he was later removed and placed on public display at the San Giovanni Rotondo.

Environmental Preservation

A body buried in a sufficiently dry environment may be well preserved for decades. This was observed in the case for murdered civil rights activist Medgar Evers, who was found to be almost perfectly preserved over 30 years after his death, permitting an accurate autopsy when the case of his murder was re-opened in the 1990s.

Bodies submerged in a peat bog may become naturally "embalmed", arresting decomposition and resulting in a preserved specimen known as a bog body. The time for an embalmed body to be reduced to a skeleton varies greatly. Even when a body is decomposed, embalming treatment can still be achieved (the arterial system decays more slowly) but would not restore a natural appearance without extensive reconstruction and cosmetic work, and is largely used to control the foul odors due to decomposition.

An animal can be preserved almost perfectly, for millions of years in a resin such as amber.

There are some examples where bodies have been inexplicably preserved (with no human intervention) for decades or centuries and appear almost the same as when they died. In some religious groups, this is known as incorruptibility. It is not known whether or for how long a body can stay free of decay without artificial preservation.

Persistent Organic Pollutant

Persistent organic pollutants (POPs) are organic compounds that are resistant to environmental degradation through chemical, biological, and photolytic processes. Because of their persistence, POPs bioaccumulate with potential significant impacts on human health and the environment. The effect of POPs on human and environmental health was discussed, with intention to eliminate or severely restrict their production, by the international community at the Stockholm Convention on Persistent Organic Pollutants in 2001.

Many POPs are currently or were in the past used as pesticides, solvents, pharmaceuticals, and industrial chemicals. Although some POPs arise naturally, for example volcanoes and various bio-synthetic pathways, most are man-made via total synthesis.

Consequences of Persistence

POPs typically are halogenated organic compounds and as such exhibit high lip-id solubility. For this reason, they bioaccumulate in fatty tissues. Halogenated compounds also exhibit great stability reflecting the nonreactivity of C-Cl bonds toward hydrolysis and photolytic degradation. The stability and lipophilicity of organic compounds often correlates with their halo-gen content, thus polyhalogenated organic compounds are of particular concern. They exert their negative effects on the environment through two processes, long range transport, which allows them to travel far from their source, and bioaccumulation, which reconcentrates these chemical compounds to potentially dangerous levels. Compounds that make up POPs are also classed as PBTs (Persistent, Bioaccumulative and Toxic) or TOMPs (Toxic Organic Micro Pollutants).

Long-Range Transport

POPs enter the gas phase under certain environmental temperatures and volatize from soils, vege-tation, and bodies of water into the atmosphere, resisting breakdown reactions in the air, to travel long distances before being re-deposited. This results in accumulation of POPs in areas far from where they were used or emitted, specifically environments where POPs have never been intro-duced such as Antarctica, and the Arctic circle. POPs can be present as vapors in the atmosphere or bound to the surface of solid particles. POPs have low solubility in water but are easily captured by solid particles, and are soluble in organic fluids (oils, fats, and liquid fuels). POPs are not eas-ily degraded in the environment due to their stability and low decomposition rates. Due to this capacity for long-range transport, POP environmental contamination is extensive, even in areas where POPs have never been used, and will remain in these environments years after restrictions implemented due to their resistance to degradation.

Bioaccumulation

Bioaccumulation of POPs is typically associated with the compounds high lipid solubility and abil-ity to accumulate in the fatty tissues of living organisms for long periods of time. Persistent chem-icals tend to have higher concentrations and are eliminated more slowly. Dietary accumulation or bioaccumulation is another hallmark characteristic of POPs, as POPs move up the food chain, they increase in concentration as they are processed and metabolized in certain tissues of organisms. The natural capacity for animals gastrointestinal tract concentrate ingested chemicals, along with poorly metabolized and hydrophobic nature of POPs makes such compounds highly susceptible to bioaccumulation. Thus POPs not only persist in the environment, but also as they are taken in by animals they bioaccumulate, increasing their concentration and toxicity in the environment.

Stockholm Convention on Persistent Organic Pollutants

The Stockholm Convention was adopted and put into practice by the United Nations Environ-ment Programme (UNEP) on May 22, 2001. The UNEP decided that POP regulation needed to be addressed globally for the future. The purpose statement of the agreement is "to protect human health and the environment from persistent organic pollutants." As of 2014, there are 179 countries in compliance with the Stockholm convention. The convention and its participants have recognized the potential human and environmental toxicity of POPs. They recognize that POPs have the potential for long range transport and bioaccumulation and biomagnification. The convention seeks to study and then judge whether or not a number of chemicals that have been developed with advances in technology and science can be categorized as POPs or not. The initial meeting in 2001 made a preliminary list, termed the "dirty dozen," of chemicals that are classified as POPs. As of 2014, the United States of America has signed the Stockholm Convention but has not ratified it. There are a handful of other countries that have not ratified the convention but most countries in the world have ratified the convention.

State parties to the Stockholm Convention on Persistent Organic Pollutants

Compounds on the Stockholm Convention List

In May 1995, the United Nations Environment Programme Governing Council investigated POPs. Initially the Convention recognized only twelve POPs for their adverse effects on human health and the environment, placing a global ban on these particularly harmful and toxic compounds and requiring its parties to take measures to eliminate or reduce the release of POPs in the environment.

1. Aldrin, an insecticide used in soils to kill termites, grasshoppers, Western corn rootworm, and others, is also known to kill birds, fish, and humans. Humans are primarily exposed to aldrin through dairy products and animal meats.

2. Chlordane, an insecticide used to control termites and on a range of agricultural crops, is known to be lethal in various species of birds, including mallard ducks, bobwhite quail, and pink shrimp; it is a chemical that remains in the soil with a reported half-life of one year. Chlordane has been postulated to affect the human immune system and is classified as a possible human carcinogen. Chlordane air pollution is believed the primary route of humane exposure.

3. Dieldrin, a pesticide used to control termites, textile pests, insect-borne diseases and insects living in agricultural soils. In soil and insects, aldrin can be oxidized, resulting in rapid conversion to dieldrin. Dieldrin's half-life is approximately five years. Dieldrin is highly toxic to fish and other aquatic animals, particularly frogs, whose embryos can develop spinal deformities after exposure to low levels. Dieldrin has been linked to Parkinson's dis-

ease, breast cancer, and classified as immunotoxic, neurotoxic, with endocrine disrupting capacity. Dieldrin residues have been found in air, water, soil, fish, birds, and mammals. Human exposure to dieldrin primarily derives from food.

4. Endrin, an insecticide sprayed on the leaves of crops, and used to control rodents. Animals can metabolize endrin, so fatty tissue accumulation is not an issue, however the chemical has a long half-life in soil for up to 12 years. Endrin is highly toxic to aquatic animals and humans as a neurotoxin. Human exposure results primarily through food.

5. Heptachlor, a pesticide primarily used to kill soil insects and termites, along with cotton insects, grasshoppers, other crop pests, and malaria-carrying mosquitoes. Heptachlor, even at every low doses has been associated with the decline of several wild bird populations – Canada geese and American kestrels. In laboratory tests have shown high-dose heptachlor as lethal, with adverse behavioral changes and reduced reproductive success at low-doses, and is classified as a possible human carcinogen. Human exposure primarily results from food.

6. Hexachlorobenzene (HCB), was first introduced in 1945–1959 to treat seeds because it can kill fungi on food crops. HCB-treated seed grain consumption is associated with photosensitive skin lesions, colic, debilitation, and a metabolic disorder called porphyria turcica, which can be lethal. Mothers who pass HCB to their infants through the placenta and breast milk had limited reproductive success including infant death. Human exposure is primarily from food.

7. Mirex, an insecticide used against ants and termites or as a flame retardant in plastics, rubber, and electrical goods. Mirex is one of the most stable and persistent pesticides, with a half-life of up to 10 years. Mirex is toxic to several plant, fish and crustacean species, with suggested carcinogenic capacity in humans. Humans are exposed primarily through animal meat, fish, and wild game.

8. Toxaphene, an insecticide used on cotton, cereal, grain, fruits, nuts, and vegetables, as well as for tick and mite control in livestock. Widespread toxaphene use in the US and chemical persistence, with a half-life of up to 12 years in soil, results in residual toxaphene in the environment. Toxaphene is highly toxic to fish, inducing dramatic weight loss and reduced egg viability. Human exposure primarily results from food. While human toxicity to direct toxaphene exposure is low, the compound is classified as a possible human carcinogen.

9. Polychlorinated biphenyls (PCBs), used as heat exchange fluids, in electrical transformers, and capacitors, and as additives in paint, carbonless copy paper, and plastics. Persistence varies with degree of halogenation, an estimated half-life of 10 years. PCBs are toxic to fish at high doses, and associated with spawning failure at low doses. Human exposure occurs through food, and is associated with reproductive failure and immune suppression. Immediate effects of PCB exposure include pigmentation of nails and mucous membranes and swelling of the eyelids, along with fatigue, nausea, and vomiting. Effects are transgenerational, as the chemical can persist in a mother's body for up to 7 years, resulting in developmental delays and behavioral problems in her children. Food contamination has led to large scale PCB exposure.

10. Dichlorodiphenyltrichloroethane (DDT) is probably the most infamous POP. It was widely used as insecticide during WWII to protect against malaria and typhus. After the war, DDT was used as an agricultural insecticide. In 1962, the American biologist Rachel Carson published Silent Spring, describing the impact of DDT spraying on the US environment and human health. DDT's persistence in the soil for up to 10–15 years after application has resulted in widespread and persistent DDT residues throughout the world including the arctic, even though it has been banned or severely restricted in most of the world. DDT is toxic to many organisms including birds where it is detrimental to reproduction due to eggshell thinning. DDT can be detected in foods from all over the world and food-borne DDT remains the greatest source of human exposure. Short-term acute effects of DDT on humans are limited, however long-term exposure has been associated with chronic health effects including increased risk of cancer and diabetes, reduced reproductive success, and neurological disease.

11. Dioxins are unintentional by-products of high-temperature processes, such as incomplete combustion and pesticide production. Dioxins are typically emitted from the burning of hospital waste, municipal waste, and hazardous waste, along with automobile emissions, peat, coal, and wood. Dioxins have been associated with several adverse effects in humans, including immune and enzyme disorders, chloracne, and are classified as a possible human carcinogen. In laboratory studies of dioxin effects an increase in birth defects and still-births, and lethal exposure have been associated with the substances. Food, particularly from animals, is the principal source of human exposure to dioxins.

12. Polychlorinated dibenzofurans are by-products of high-temperature processes, such as incomplete combustion after waste incineration or in automobiles, pesticide production, and polychlorinated biphenyl production. Structurally similar to dioxins, the two compounds share toxic effects. Furans persist in the environment and classified as possible human carcinogens. Human exposure to furans primarily results from food, particularly animal products.

New POPs on the Stockholm Convention List

Since 2001, this list has been expanded to include some polycyclic aromatic hydrocarbons (PAHs), brominated flame retardants, and other compounds. Additions to the initial 2001 Stockholm Convention list are as following POPs:

- Chlordecone, a synthetic chlorinated organic compound,is primarily used as an agricultural pesticide, related to DDT and Mirex. Chlordecone is toxic to aquatic organisms, and classified as a possible human carcinogen. Many countries have banned chlordecone sale and use, or intend to phase out stockpiles and wastes.

- α-Hexachlorocyclohexane (α-HCH) and β-Hexachlorocyclohexane (β-HCH) are insecticides as well as by-products in the production of lindane. Large stockpiles of HCH isomers exist in the environment. α-HCH and β-HCH are highly persistent in the water of colder regions. α-HCH and β-HCH has been linked Parkinson's and Alzheimer's disease.

- Hexabromodiphenyl ether (hexaBDE) and heptabromodiphenyl ether (heptaBDE) are main components of commercial octabromodiphenyl ether (octaBDE). Commercial octaB-

DE is highly persistent in the environment, whose only degradation pathway is through debromination and the production of bromodiphenyl ethers, which can increase toxicity.

- Lindane (γ-hexachlorocyclohexane), a pesticide used as a broad spectrum insecticide for seed, soil, leaf, tree and wood treatment, and against ectoparasites in animals and humans (head lice and scabies). Lindane rapidly bioconcentrates. It is immunotoxic, neurotoxic, carcinogenic, linked to liver and kidney damage as well as adverse reproductive and developmental effects in laboratory animals and aquatic organisms. Production of lindane unintentionally produces two other POPs α-HCH and β-HCH.

- Pentachlorobenzene (PeCB), is a pesticide and unintentional byproduct. PeCB has also been used in PCB products, dyestuff carriers, as a fungicide, a flame retardant, and a chemical intermediate. PeCB is moderately toxic to humane, while highly toxic to aquatic organisms.

- Tetrabromodiphenyl ether (tetraBDE) and pentabromodiphenyl ether (pentaBDE) are industrial chemicals and the main components of commercial pentabromodiphenyl ether (pentaBDE). PentaBDE has been detected in humans in all regions of the world.

- Perfluorooctanesulfonic acid (PFOS) and its salts are used in the production of fluoropolymers. PFOS and related compounds are extremely persistent, bioaccumulating and biomagnifying. The negative effects of trace levels of PFOS have not been established.

- Endosulfans are insecticides to control pests on crops such coffee, cotton, rice and sorghum and soybeans, tsetse flies, ectoparasites of cattle. They are used as a wood preservative. Global use and manufacturing of endosulfan has been banned under the Stockholm convention in 2011, although many countries had previously banned or introduced phase-outs of the chemical when the ban was announced. Toxic to humans and aquatic and terrestrial organisms, linked to congenital physical disorders, mental retardation, and death. Endosulfans' negative health effects are primarily liked to its endocrine disrupting capacity acting as an antiandrogen.

- Hexabromocyclododecane (HBCD) is a brominated flame retardant primarily used in thermal insulation in the building industry. HBCD is persistent, toxic and ecotoxic, with bioaccumulative and long-range transport properties.

Additive and Synergistic Effects

Evaluation of the effects of POPs on health is very challenging in the laboratory setting. For example, for organisms exposed to a mixture of POPs, the effects are assumed to be additive. Mixtures of POPs can in principle produce synergistic effects. With synergistic effects, the toxicity of each compound is enhanced (or depressed) by the presence of other compounds in the mixture. When put together, the effects can far exceed the approximated additive effects of the POP compound mixture.

Health Effects

POP exposure may cause developmental defects, chronic illnesses, and death. Some are carcinogens per IARC, possibly including breast cancer. Many POPs are capable of endocrine disruption within the reproductive system, the central nervous system, or the immune system. People and animals are exposed to POPs mostly through their diet, occupationally, or while growing in the womb. For humans not exposed to POPs through accidental or occupational means, over 90% of exposure comes from animal product foods due to bioaccumulation in fat tissues and bioaccumulate through the food chain. In general, POP serum levels increase with age and tend to be higher in females than males.

Studies have investigated the correlation between low level exposure of POPs and various diseases. In order to assess disease risk due to POPs in a particular location, government agencies may produce a human health risk assessment which takes into account the pollutants' bioavailability and their dose-response relationships.

Endocrine Disruption

The majority of POPs are known to disrupt normal functioning of the endocrine system, for example all of the dirty dozen are endocrine disruptors. Low level exposure to POPs during critical developmental periods of fetus, newborn and child can have a lasting effect throughout its lifespan. A 2002 study synthesizes data on endocrine disruption and health complications from exposure to POPs during critical developmental stages in an organism's lifespan. The study aimed to answer the question whether or not chronic, low level exposure to POPs can have a health impact on the endocrine system and development of organisms from different species. The study found that exposure of POPs during a critical developmental time frame can produce a permanent changes in the organisms path of development. Exposure of POPs during non-critical developmental time frames may not lead to detectable diseases and health complications later in their life. In wildlife, the critical development time frames are in utero, in ovo, and during reproductive periods. In humans, the critical development timeframe is during fetal development.

Reproductive System

The same study in 2002 with evidence of a link from POPs to endocrine disruption also linked low dose exposure of POPs to reproductive health effects. The study stated that POP exposure can lead to negative health effects especially in the male reproductive system, such as decreased sperm quality and quantity, altered sex ratio and early puberty onset. For females exposed to POPs, altered reproductive tissues and pregnancy outcomes as well as endometriosis have been reported.

Exposure During Pregnancy

POP exposure during pregnancy is of particular concern to the developing fetus.

Transport Across the Placenta

A study about the transfer of POPs (14 organochlorine pesticides, 7 polychlorinated biphenyls and 14 polybrominated diphenyl ethers (PBDEs)) from Spanish mothers to their unborn fetus found that POP concentrations in serum from the mother were higher than from the umbilical cord and 50 placentas. Because transfer of the POPs from mother to fetus did not correspond with passive lipid-associated diffusion, authors suggested that POPs are actively transported across the placenta.

Gestational Weight Gain and Newborn Head Circumference

A Greek study from 2014 investigated the link between maternal weight gain during pregnancy, their PCB-exposure level and PCB level in their newborn infants, their birth weight, gestational age, and head circumference. The birth weight and head circumference of the infants was the lower, the higher POP levels during prenatal development had been, but only if mothers had either excessive or inadequate weight gain during pregnancy. No correlation between POP exposure and gestational age was found. A 2013 case-control study conducted 2009 in Indian mothers and their offspring showed prenatal exposure of two types of organochlorine pesticides (HCH, DDT and DDE) impaired the growth of the fetus, reduced the birth weight, length, head circumference and chest circumference.

Cardiovascular Disease and Cancer

POPs are lipophilic environmental toxins. They are often found in lipoproteins of organisms. A study published in 2014 found an association between the concentration of POPs in lipoproteins and the occurrence of cardiovascular disease and various cancers in human beings. The higher the concentration of POPs found in lipoproteins, the higher the occurrence of cardiovascular disease and cancer. Highly chlorinated polychlorinated biphenyls are specifically found in high concentrations in lipoproteins. Cardiovascular disease is shown to be more associated with higher concentrations of POPs in high density lipoproteins and cancer is shown to be more associated with higher concentrations of POPs in low density lipoproteins and very low density lipoproteins.

Obesity

There have been many recent studies assessing the connection between serum POP levels in individuals and instances of obesity. A study released in 2011 found correlations between different POPs and obesity occurrence in individuals tested. The statistically significant findings from the study show that there is actually a negative correlation between various PCB congener serum levels and obesity in individuals tested. The study also showed a positive correlation between beta-hexachlorocyclohexane and various dioxin serum levels and obesity in individuals tested. Obesity was determined using the Body Mass Index (BMI). One proposed explanation in the study is that PCBs are very lipophilic, therefore they are easily stored and captured in the fat deposits in human beings. Obese individuals have higher amounts of fat deposits in their body, and thus more PCBs could be captured in the fat deposits leading to less PCBs circulating in blood serum. The study provides evidence demonstrating that the correlation between POP serum levels and obesity occurrence is more complicated than previously expected. The same study also noted a strong positive correlation between serum POP levels and age in all individuals in the experiment.

Diabetes

A study published in 2006 revealed a positive correlation between POP serum levels and type II diabetes in individuals, after other variables, such as age, sex, race, and socioeconomic status were adjusted for. The correlation proved stronger in younger, Mexican American, and obese individuals. Individuals exposed to low doses of POPs throughout their lifetime had a higher chance for developing diabetes than individuals exposed to high concentrations of POPs for a short amount of time.

POPs in Urban Areas and Indoor Environments

Traditionally it was thought that human exposure to POPs occurred primarily through food, however indoor pollution patterns that characterize certain POPs have challenged this notion. Recent studies of indoor dust and air have implicated indoor environments as a major sources for human exposure via inhalation and ingestion. Furthermore, significant indoor POP pollution must be a major route of human POP exposure, considering the modern trend in spending larger proportions of life indoor. Several studies have shown that indoor (air and dust) POP levels to exceed outdoor (air and soil) POP concentrations.

Control and Removal of POPs in the Environment

Current studies aimed at minimizing POPs in the environment are investigating their behavior in photo catalytic oxidation reactions. POPs that are found in humans and in aquatic environments the most are the main subjects of these experiments. Aromatic and aliphatic degradation products have been identified in these reactions. Photochemical degradation is negligible compared to photocatalytic degradation. However, proper removal techniques of POPs from the environment are still unclear, due to fear that more toxic byproducts may result from uninvestigated degradation techniques. Current efforts are more focused on banning the use and production of POPs worldwide rather than removal of POPs.

References

• Zehnder, Alexander J. B. (1978). "Ecology of methane formation". In Mitchell, Ralph. Water pollution microbiology 2. New York: Wiley. pp. 349–376. ISBN 978-0-471-01902-2.

• Jenkins, J.C. (2005). The Humanure Handbook: A Guide to Composting Human Manure. Grove City, PA: Joseph Jenkins, Inc.; 3rd edition. p. 255. ISBN 978-0-9644258-3-5. Retrieved April 2011. Check date values in: |access-date= (help)

• Hemenway, Toby (2009). Gaia's Garden: A Guide to Home-Scale Permaculture. Chelsea Green Publishing. pp. 84-85. ISBN 978-1-60358-029-8.

• Forbes, S.L. (2008). "Decomposition Chemistry in a Burial Environment". In M. Tibbett; D.O. Carter. Soil Analysis in Forensic Taphonomy. CRC Press. pp. 203–223. ISBN 1-4200-6991-8.

• Carter D.O.; Tibbett M. (2008). "Cadaver Decomposition and Soil: Processes". In M. Tibbett; D.O. Carter. Soil Analysis in Forensic Taphonomy. CRC Press. pp. 29–51. ISBN 1-4200-6991-8.

• Pinheiro, J. (2006). "Decay Process of a Cadaver". In A. Schmidt; E. Cumha; J. Pinheiro. Forensic Anthropology and Medicine. Humana Press. pp. 85–116. ISBN 1-58829-824-8.

• Schmitt, A.; Cunha, E.; Pinheiro, J. (2006). Forensic Anthropology and Medicine: Complementary Sciences From Recovery to Cause of Death. Humana Press. p. 464. ISBN 1-58829-824-8.

• Haglund, WD.; Sorg, MH. (1996). Forensic Taphonomy: The Postmortem Fate of Human Remains. CRC Press. p. 636. ISBN 0-8493-9434-1.

• Quigley, C. (1998). Modern Mummies: The Preservation of the Human Body in the Twentieth Century. McFarland. pp. 213–214. ISBN 0-7864-0492-2.

• Martin & Gershuny eds. (1992). The Rodale Book of Composting: Easy Methods for Every Gardener. Rodale Press. Retrieved 26 October 2015.

• Astoviza, Malena J. (15 April 2014). "Evaluación de la distribución de contaminantes orgánicos persistentes (COPs) en aire en la zona de la cuenca del Plata mediante muestreadores pasivos artificiales" (in Spanish): 160.

Retrieved 16 April 2014.

- Vizcaino, E; Grimalt JO; Fernández-Somoano A; Tardon A (2014). "Transport of persistent organic pollutants across the human placenta". Environ Int. 65: 107–115. doi:10.1016/j.envint.2014.01.004. PMID 24486968.

- Tilley, E., Ulrich, L., Lüthi, C., Reymond, Ph., Zurbrügg, C. (2014) Compendium of Sanitation Systems and Technologies - (2nd Revised Edition). Swiss Federal Institute of Aquatic Science and Technology (Eawag), Duebendorf, Switzerland.

3

Biodegradable Plastic: An Overview

Plastic that can be decomposed by bacteria is known as biodegradable plastic. Plastic, by nature, can damage the environment and biodegradable plastic is an answer and a cure for that. This chapter introduces the reader to fascinating concepts like bioplastic, polycaprolactone, polyvinyl alcohol and polyproylene carbonate.

Biodegradable Plastic

Biodegradable plastics are plastics that decompose by the action of living organisms, usually bacteria.

Utensils made from biodegradable plastic

Two basic classes of biodegradable plastics exist: Bioplastics, whose components are derived from renewable raw materials, and plastics made from petrochemicals containing biodegradable additives which enhance biodegradation.

Examples of Biodegradable Plastics

- While aromatic polyesters are almost totally resistant to microbial attack, most aliphatic polyesters are biodegradable due to their potentially hydrolysable ester bonds:
    - Naturally Produced: Polyhydroxyalkanoates (PHAs) like the poly-3-hydroxybutyrate (PHB), polyhydroxyvalerate (PHV) and polyhydroxyhexanoate (PHH);

     o   Renewable Resource: Polylactic acid (PLA);

     o   Synthetic: Polybutylene succinate (PBS), polycaprolactone (PCL)...

- Polyanhydrides
- Polyvinyl alcohol
- Most of the starch derivatives
- Cellulose esters like cellulose acetate and nitrocellulose and their derivatives (celluloid).
- Polyethylene terephthalate
- Enhanced biodegradable plastic with additives.

Development of biodegradable containers

Controversy

Many people confuse "biodegradable" with "compostable". "Biodegradable" broadly means that an object can be biologically broken down, while "compostable" typically specifies that such a process will result in compost, or humus. Many plastic manufacturers throughout Canada and the US have released products indicated as being compostable. However this claim is debatable, if the manufacturer was minimally conforming to the now-withdrawn American Society for Testing and Materials standard definition of the word, as it applies to plastics:

"that which is capable of undergoing biological decomposition in a compost site such that the material is not visually distinguishable and breaks down into carbon dioxide, water, inorganic compounds and biomass at a rate consistent with known compostable materials." (ASTM D 6002)

There is a major discrepancy between this definition and what one would expect from a backyard composting operation. With the inclusion of "inorganic compounds", the above definition allows that the end product might not be humus, an organic substance. The only criterion the ASTM standard definition *did* outline is that a compostable plastic has to become "not visually distinguishable" at the same rate as something that has already been established as being compostable under the traditional definition.

Withdrawal of ASTM D 6002

In January 2011, the ASTM withdrew standard ASTM D 6002, which many plastic manufacturers had been referencing to attain credibility in labelling their products as compostable. The withdrawn description was as follows:

"This guide covered suggested criteria, procedures, and a general approach to establish the compostability of environmentally degradable plastics."

As of 2014, the ASTM has yet to replace this standard.

Advantages and Disadvantages

Under proper conditions, some biodegradable plastics can degrade to the point where microorganisms can completely metabolise them to carbon dioxide (and water). For example, starch-based bioplastics produced from sustainable farming methods could be almost carbon neutral.

There are allegations that "Oxo Biodegradable (OBD)" plastic bags may release metals, and may require a great deal of time to degrade in certain circumstances and that OBD plastics may produce tiny fragments of plastic that do not continue to degrade at any appreciable rate regardless of the environment. The response of the Oxo-biodegradable Plastics Association (www.biodeg.org) is that OBD plastics do not contain metals. They contain salts of metals, which are not prohibited by legislation and are in fact necessary as trace-elements in the human diet. Oxo-biodegradation of polymer material has been studied in depth at the Technical Research Institute of Sweden and the Swedish University of Agricultural Sciences. A peer-reviewed report of the work was published in Vol 96 of the journal of Polymer Degradation & Stability (2011) at page 919-928, which shows 91% biodegradation in a soil environment within 24 months, when tested in accordance with ISO 17556.

Environmental Benefits

There is much debate about the total carbon, fossil fuel and water usage in manufacturing bioplastics from natural materials and whether they are a negative impact to human food supply. To make 1 kg (2.2 lb) of polylactic acid, the most common commercially available compostable plastic, 2.65 kg (5.8 lb) of corn is required. Since 270 million tonnes of plastic are made every year, replacing conventional plastic with corn-derived polylactic acid would remove 715.5 million tonnes from the world's food supply, at a time when global warming is reducing tropical farm productivity. "Although U.S. corn is a highly productive crop, with typical yields between 140 and 160 bushels per acre, the resulting delivery of food by the corn system is far lower. Today's corn crop is mainly used for biofuels (roughly 40 percent of U.S. corn is used for ethanol) and as animal feed (roughly 36 percent of U.S. corn, plus distillers grains left over from ethanol production, is fed to cattle, pigs and chickens). Much of the rest is exported. Only a tiny fraction of the national corn crop is directly used for food for Americans, much of that for high-fructose corn syrup."

Traditional plastics made from non-renewable fossil fuels lock up much of the carbon in the plastic, as opposed to being burned in the processing of the plastic. The carbon is permanently trapped inside the plastic lattice, and is rarely recycled, if one neglects to include the diesel, pesticides, and fertilizers used to grow the food turned into plastic.

There is concern that another greenhouse gas, methane, might be released when any biodegradable material, including truly biodegradable plastics, degrades in an anaerobic landfill environment. Methane production from 594 managed landfill environments is captured and used for energy;some landfills burn this off through a process called flaring to reduce the release of methane into the environment. In the US, most landfilled materials today go into landfills where they capture the methane biogas for use in clean, inexpensive energy. Incinerating non-biodegradable plastics will release carbon dioxide as well. Disposing of non-biodegradable plastics made from natural materials in anaerobic (landfill) environments will result in the plastic lasting for hundreds of years.

Bacteria have developed the ability to degrade plastics. This has already happened with nylon: two types of nylon eating bacteria, *Flavobacteria* and *Pseudomonas*, were found in 1975 to possess enzymes (nylonase) capable of breaking down nylon. While not a solution to the disposal problem, it is likely that bacteria have developed the ability to consume hydrocarbons. In 2008, a 16-year-old boy reportedly isolated two plastic-consuming bacteria.

Environmental Concerns and Benefits

According to a 2010 EPA report, 12.4%, or 31 million tons, of all municipal solid waste (MSW) is plastic. 8.2% of that, or 2.55 million tons, were recovered. That is significantly lower than the average recovery percentage of 34.1%.

Much of the reason for disappointing plastics recycling goals is that conventional plastics are often commingled with organic wastes (food scraps, wet paper, and liquids), making it difficult and impractical to recycle the underlying polymer without expensive cleaning and sanitizing procedures.

On the other hand, composting of these mixed organics (food scraps, yard trimmings, and wet, non-recyclable paper) is a potential strategy for recovering large quantities of waste and dramatically increasing community recycling goals. Food scraps and wet, non-recyclable paper comprise 50 million tons of municipal solid waste. Biodegradable plastics can replace the non-degradable plastics in these waste streams, making municipal composting a significant tool to divert large amounts of otherwise nonrecoverable waste from landfills.

Compostable plastics combine the utility of plastics (lightweight, resistance, relative low cost) with the ability to completely and fully compost in an industrial compost facility. Rather than worrying about recycling a relatively small quantity of commingled plastics, proponents argue that certified biodegradable plastics can be readily commingled with other organic wastes, thereby enabling composting of a much larger position of nonrecoverable solid waste. Commercial composting for all mixed organics then becomes commercially viable and economically sustainable. More municipalities can divert significant quantities of waste from overburdened landfills since the entire waste stream is now biodegradable and therefore easier to process. This move away from the use of landfills may help alleviate the issue of plastic pollution.

The use of biodegradable plastics, therefore, is seen as enabling the complete recovery of large quantities of municipal sold waste (via aerobic composting) that have heretofore been unrecoverable by other means except land filling or incineration.

Energy Costs for Production

Various researchers have undertaken extensive life cycle assessments of biodegradable polymers to determine whether these materials are more energy efficient than polymers made by conventional fossil fuel-based means. Research done by Gerngross, *et al.* estimates that the fossil fuel energy required to produce a kilogram of polyhydroxyalkanoate (PHA) is 50.4 MJ/kg, which coincides with another estimate by Akiyama, *et al.*, who estimate a value between 50-59 MJ/kg. This information does not take into account the feedstock energy, which can be obtained from non-fossil fuel based methods. Polylactide (PLA) was estimated to have a fossil fuel energy cost of 54-56.7 from two sources, but recent developments in the commercial production of PLA by NatureWorks has eliminated some dependence of fossil fuel-based energy by supplanting it with wind power and biomass-driven strategies. They report making a kilogram of PLA with only 27.2 MJ of fossil fuel-based energy and anticipate that this number will drop to 16.6 MJ/kg in their next generation plants. In contrast, polypropylene and high-density polyethylene require 85.9 and 73.7 MJ/kg, respectively, but these values include the embedded energy of the feedstock because it is based on fossil fuel.

Gerngross reports a 2.65 kg total fossil fuel energy equivalent (FFE) required to produce a single kilogram of PHA, while polyethylene only requires 2.2 kg FFE. Gerngross assesses that the decision to proceed forward with any biodegradable polymer alternative will need to take into account the priorities of society with regard to energy, environment, and economic cost.

Furthermore, it is important to realize the youth of alternative technologies. Technology to produce PHA, for instance, is still in development today, and energy consumption can be further reduced by eliminating the fermentation step, or by utilizing food waste as feedstock. The use of alternative crops other than corn, such as sugar cane from Brazil, are expected to lower energy requirements. For instance, manufacturing of PHAs by fermentation in Brazil enjoys a favorable energy consumption scheme where bagasse is used as source of renewable energy.

Many biodegradable polymers that come from renewable resources (i.e. starch-based, PHA, PLA) also compete with food production, as the primary feedstock is currently corn. For the US to meet its current output of plastics production with BPs, it would require 1.62 square meters per kilogram produced. While this space requirement could be feasible, it is always important to consider how much impact this large scale production could have on food prices and the opportunity cost of using land in this fashion versus alternatives.

Regulation

United States

In terms of ASTM industrial standard definitions, the U.S.Federal Trade Commission and the U.S. EPA set standards for biodegradability. ASTM International defines methods to test for biodegradable plastic, both anaerobically and aerobically, as well as in marine environments. The specific subcommittee responsibility for overseeing these standards falls on the Committee D20.96 on Environmentally Degradable Plastics and Bio based Products. The current ASTM standards are defined as standard specifications and standard test methods. Standard specifications create a pass or fail scenario whereas standard test methods identify the specific testing parameters for facilitating specific time frames and toxicity of biodegradable tests on plastics.

Two testing methods are defined for anaerobic environments: (1) ASTM D5511-12 and (2) ASTM D5526 - 12 Standard Test Method for Determining Anaerobic Biodegradation of Plastic Materials Under Accelerated Landfill Conditions, Both of these tests are used for the ISO DIS 15985 on determining anaerobic biodegradation of plastic materials.

Bioplastic

Bioplastics are plastics derived from renewable biomass sources, such as vegetable fats and oils, corn starch, or microbiota. Bioplastic can be made from agricultural by-products and also from used plastic bottles and other containers using microorganisms. Common plastics, such as fossil-fuel plastics (also called petrobased polymers), are derived from petroleum or natural gas. Production of such plastics tends to require more fossil fuels and to produce more greenhouse gases than the production of biobased polymers (bioplastics). Some, but not all, bioplastics are designed to biodegrade. Biodegradable bioplastics can break down in either anaerobic or aerobic environments, depending on how they are manufactured. Bioplastics can be composed of starches, cellulose, biopolymers, and a variety of other materials.

IUPAC Definition

Biobased polymer derived from the *biomass* or issued from monomers derivedfrom the biomass and which, at some stage in its processing into finishedproducts, can be shaped by flow.

Packaging peanuts made from bioplastics (thermoplastic starch)

Plastics packaging made from bioplastics and other biodegradable plastics

Applications

Bioplastics are used for disposable items, such as packaging, crockery, cutlery, pots, bowls, and straws. They are also often used for bags, trays, fruit and vegetable containers and blister foils, egg cartons, meat packaging, vegetables, and bottling for soft drinks and dairy products.

Flower wrapping made of PLA-blend bio-flex

These plastics are also used in non-disposable applications including mobile phone casings, carpet fibers, insulation car interiors, fuel lines, and plastic piping. New electroactive bioplastics are being developed that can be used to carry electric current. In these areas, the goal is not biodegradability, but to create items from sustainable resources.

Medical implants made of PLA (polylactic acid), which dissolve in the body, can save patients a second operation. Compostable mulch films can also be produced from starch polymers and used in agriculture. These films do not have to be collected after use on farm fields.

Biopolymers are available as coatings for paper rather than the more common petrochemical coatings.

Bioplastic Types

Starch-Based Plastics

Thermoplastic starch currently represents the most widely used bioplastic, constituting about 50 percent of the bioplastics market. Simple starch bioplastic can be made at home. Pure starch is able to absorb humidity, and is thus a suitable material for the production of drug capsules by the pharmaceutical sector. Flexibiliser and plasticiser such as sorbitol and glycerine can also be added so the starch can also be processed thermo-plastically. The characteristics of the resulting bioplastic (also called "thermo-plastical starch") can be tailored to specific needs by adjusting the amounts of these additives.

Starch-based bioplastics are often blended with biodegradable polyesters to produce starch / polycaprolactone or starch/Ecoflex (polybutylene adipate-co-terephthalate produced by BASF). blends. These blends are used for industrial applications and are also compostable. Other producers, such as Roquette, have developed other starch/polyolefin blends. These blends are not biodegradable, but have a lower carbon footprint than petroleum-based plastics used for the same applications.

Cellulose-Based Plastics

Cellulose bioplastics are mainly the cellulose esters, (including cellulose acetate and nitrocellulose) and their derivatives, including celluloid.

A packaging blister made from cellulose acetate, a bioplastic

Some Aliphatic Polyesters

The aliphatic biopolyesters are mainly polyhydroxyalkanoates (PHAs) like the poly-3-hydroxybutyrate (PHB), polyhydroxyvalerate (PHV) and polyhydroxyhexanoate (PHH).

Polylactic Acid (PLA)

Mulch film made of polylactic acid (PLA)-blend bio-flex

Polylactic acid (PLA) is a transparent plastic produced from corn or dextrose. Its characteristics are similar to conventional petrochemical-based mass plastics (like PET, PS or PE), and it can be processed using standard equipment that already exists for the production of some conventional plastics. PLA and PLA blends generally come in the form of granulates with various properties, and are used in the plastic processing industry for the production of films, fibers, plastic containers, cups and bottles.

Poly-3-Hydroxybutyrate (PHB)

The biopolymer poly-3-hydroxybutyrate (PHB) is a polyester produced by certain bacteria processing glucose, corn starch or wastewater. Its characteristics are similar to those of the petroplastic polypropylene. PHB production is increasing. The South American sugar industry, for example, has decided to expand PHB production to an industrial scale. PHB is distinguished primarily by its physical characteristics. It can be processed into a transparent film with a melting point higher than 130 degrees Celsius, and is biodegradable without residue.

Polyhydroxyalkanoates (PHA)

Polyhydroxyalkanoates are linear polyesters produced in nature by bacterial fermentation of sugar or lipids. They are produced by the bacteria to store carbon and energy. In industrial production, the polyester is extracted and purified from the bacteria by optimizing the conditions for the fermentation of sugar. More than 150 different monomers can be combined within this family to give materials with extremely different properties. PHA is more ductile and less elastic than other plastics, and it is also biodegradable. These plastics are being widely used in the medical industry.

Polyamide 11 (PA 11)

PA 11 is a biopolymer derived from natural oil. It is also known under the tradename Rilsan B, commercialized by Arkema. PA 11 belongs to the technical polymers family and is not biodegradable. Its properties are similar to those of PA 12, although emissions of greenhouse gases and consumption of nonrenewable resources are reduced during its production. Its thermal resistance is also superior to that of PA 12. It is used in high-performance applications like automotive fuel lines, pneumatic airbrake tubing, electrical cable antitermite sheathing, flexible oil and gas pipes, control fluid umbilicals, sports shoes, electronic device components, and catheters.

A similar plastic is Polyamide 410 (PA 410), derived 70% from castor oil, under the trade name EcoPaXX, commercialized by DSM. PA 410 is a high-performance polyamide that combines the benefits of a high melting point (approx. 250 °C), low moisture absorption and excellent resistance to various chemical substances.

Bio-Derived Polyethylene

The basic building block (monomer) of polyethylene is ethylene. Ethylene is chemically similar to, and can be derived from ethanol, which can be produced by fermentation of agricultural feedstocks such as sugar cane or corn. Bio-derived polyethylene is chemically and physically identical to traditional polyethylene – it does not biodegrade but can be recycled. Bio derivation of polyethylene can also reduce greenhouse gas emissions considerably. Brazilian chemicals group Braskem claims that using its method of producing polyethylene from sugar cane ethanol captures (removes from the environment) 2.15 tonnes of CO_2 per tonne of Green Polyethylene produced.

Green PE has the same properties, performance and application versatility as fossil-based poly-ethylene, which makes it a drop-in replacement in the plastic production chain. For these same reasons, it is also recyclable in the same recycling chain used by traditional polyethylene. Because it is part of the portfolio of high-density polyethylene (HDPE) and linear low-density polyethylene (LLDPE) products, Green PE is an option for applications in rigid and flexible packaging, closures, bags and other products.

Genetically Modified Bioplastics

Genetic modification (GM) is also a challenge for the bioplastics industry. None of the currently available bioplastics – which can be considered first generation products – require the use of GM crops, although GM corn is the standard feedstock.

Looking further ahead, some of the second generation bioplastics manufacturing technologies under development employ the "plant factory" model, using genetically modified crops or genetically modified bacteria to optimise efficiency.

Polyhydroxyurethanes

Recently, there have been a large emphasis on producing biobased and isocyanate-free polyure-thanes. One such example utilizes a spontaneous reaction between polyamines and cyclic carbon-ates to produce polyhydroxurethanes. Unlike traditional cross-linked polyurethanes, cross-linked polyhydroxyurethanes have been shown to be capable of recycling and reprocessing through dy-namic transcarbamoylation reactions.

Environmental Impact

The environmental impact of bioplastics is often debated, as there are many different metrics for "greenness" (e.g., water use, energy use, deforestation, biodegradation, etc.) and tradeoffs often exist. The debate is also complicated by the fact that many different types of bioplastics exist, each with different environmental strengths and weaknesses, so not all bioplastics can be treated as equal.

Confectionery packaging made of PLA-blend bio-flex

Bottles made from cellulose acetate biograde

The production and use of bioplastics is sometimes regarded as a more sustainable activity when compared with plastic production from petroleum (petroplastic), because it requires less fossil fuel for its production and also introduces fewer, net-new greenhouse emissions if it biodegrades. The use of bioplastics can also result in less hazardous waste than oil-derived plastics, which remain solid for hundreds of years.

Drinking straws made of PLA-blend bio-flex

Petroleum is often still used as a source of materials and energy in the production of bioplastic. Petroleum is required to power farm machinery, to irrigate crops, to produce fertilisers and pesticides, to transport crops and crop products to processing plants, to process raw materials, and ultimately to produce the bioplastic. However, it is possible to produce bioplastic using renewable energy sources and avoid the use of petroleum.

Jar made of PLA-blend bio-flex, a bioplastic

Italian bioplastic manufacturer Novamont states in its own environmental audit that producing one kilogram of its starch-based product uses 500 g of petroleum and consumes almost 80% of the energy required to produce a traditional polyethylene polymer. Environmental data from Nature-Works, the only commercial manufacturer of PLA (polylactic acid) bioplastic, says that making its plastic material delivers a fossil fuel saving of between 25 and 68 per cent compared with polyethylene, in part due to its purchasing of renewable energy certificates for its manufacturing plant.

A detailed study examining the process of manufacturing a number of common packaging items from traditional plastics and polylactic acid carried out by Franklin Associates and published by the Athena Institute shows that using bioplastic has a lower environmental impact for some products, and a higher environmental impact for others. This study, however, does not factor in the end-of-life environmental impact of these products, including possible methane emissions from landfills due to biodegradable plastics.

While production of most bioplastics results in reduced carbon dioxide emissions compared to traditional alternatives, there is concern that the creation of a global bioeconomy required to produce bioplastic in large quantities could contribute to an accelerated rate of deforestation and soil erosion, and could adversely affect water supplies. Careful management of a global bioeconomy would be required.

Other studies showed that bioplastics result in a 42% reduction in carbon footprint.

On October 21, 2010, a group of scientists reported that corn-based plastic ranked higher in environmental defects than the main products it replaces, such as HDPE, LDPE and PP. In the study, the production of corn-based plastics created more acidification, carcinogens, ecotoxicity, eutrophication, ozone depletion, respiratory effects and smog than the synthetic-based plastics they replaced. However the study also concluded that biopolymers trumped the other plastics for biodegradability, low toxicity, and use of renewable resources.

The American Carbon Registry has also released reports of nitrous oxide caused from corn growing which is 310 times more potent than CO_2. Pesticides are also used in growing corn-based plastic.

Bioplastics and Biodegradation

Packaging air pillow made of PLA-blend bio-flex

The terminology used in the bioplastics sector is sometimes misleading. Most in the industry use the term bioplastic to mean a plastic produced from a biological source. All (bio- and petroleum-based) plastics are technically biodegradable, meaning they can be degraded by microbes under suitable conditions. However, many degrade so slowly that they are considered non-biodegradable. Some petrochemical-based plastics are considered biodegradable, and may be used as an additive to improve the performance of commercial bioplastics. Non-biodegradable bioplastics are referred to as durable. The biodegradability of bioplastics depends on temperature, polymer stability, and available oxygen content. The European standard EN13432, published by the International Organization for Standardization, defines how quickly and to what extent a plastic must be degraded under the tightly controlled and aggressive conditions (at or above 140 °F) of an industrial composting unit for it to be considered biodegradable. This standard is recognized in many countries, including all of Europe, Japan and the US. However, it applies only to industrial composting units and does not set out a standard for home composting. Most bioplastics (e.g. PH) only biodegrade quickly in industrial composting units. These materials do not biodegrade quickly in ordinary compost piles or in the soil/water. Starch-based bioplastics are an exception, and will biodegrade in normal composting conditions.

The term "biodegradable plastic" has also been used by producers of specially modified petrochemical-based plastics that appear to biodegrade. Biodegradable plastic bag manufacturers that have misrepresented their product's biodegradability may now face legal action in the US state of California for the misleading use of the terms biodegradable or compostable. Traditional plastics such as polyethylene are degraded by ultra-violet (UV) light and oxygen. To prevent this, process manufacturers add stabilising chemicals. However, with the addition of a degradation initiator to the plastic, it is possible to achieve a controlled UV/oxidation disintegration process. This type of plastic may be referred to as *degradable plastic* or *oxy-degradable plastic* or *photodegradable plastic* because the process is not initiated by microbial action. While some degradable plastics manufacturers argue that degraded plastic residue will be attacked by microbes, these degradable materials do not meet the requirements of the EN13432 commercial composting standard. The bioplastics industry has widely criticized oxo-biodegradable plastics, which the industry association says do not meet its requirements. Oxo-biodegradable plastics – known as "oxos" – are conventional petroleum-based products with some additives that initiate degradation. The ASTM standard for oxo-biodegradables is called the Standard Guide for Exposing and Testing Plastics that Degrade in the Environment by a Combination of Oxidation and Biodegradation (ASTM 6954). Both EN 13432 and ASTM 6400 are specifically designed for PLA and Starch based products and should not be used as a guide for oxos.

Market

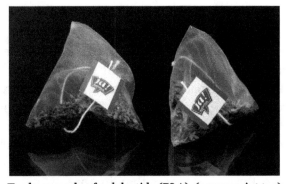

Tea bags made of polylactide (PLA), (peppermint tea)

Because of the fragmentation in the market and ambiguous definitions it is difficult to describe the total market size for bioplastics, but estimates put global production capacity at 327,000 tonnes. In contrast, global consumption of all flexible packaging is estimated at around 12.3 million tonnes.

COPA (Committee of Agricultural Organisation in the European Union) and COGEGA (General Committee for the Agricultural Cooperation in the European Union) have made an assessment of the potential of bioplastics in different sectors of the European economy:

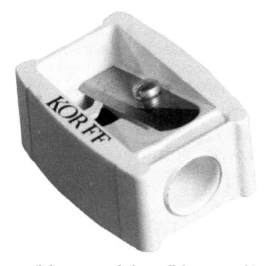

Prism pencil sharpener made from cellulose acetate biograde

- Catering products: 450,000 tonnes per year

- Organic waste bags: 100,000 tonnes per year

- Biodegradable mulch foils: 130,000 tonnes per year

- Biodegradable foils for diapers 80,000 tonnes per year

- Diapers, 100% biodegradable: 240,000 tonnes per year

- Foil packaging: 400,000 tonnes per year

- Vegetable packaging: 400,000 tonnes per year

- Tyre components: 200,000 tonnes per year

- Total: 2,000,000 tonnes per year

In the years 2000 to 2008, worldwide consumption of biodegradable plastics based on starch, sugar, and cellulose – so far the three most important raw materials – has increased by 600%. The NNFCC predicted global annual capacity would grow more than six-fold to 2.1 million tonnes by 2013. BCC Research forecasts the global market for biodegradable polymers to grow at a compound average growth rate of more than 17 percent through 2012. Even so, bioplastics will encompass a small niche of the overall plastic market, which is forecast to reach 500 billion pounds (220 million tonnes) globally by 2010. Ceresana forecasts the world market for bioplastics to reach 5.8 billion US dollars in 2021 - i.e. three times more than in 2014.

Cost

At one time bioplastics were too expensive for consideration as a replacement for petroleum-based plastics. However, the lower temperatures needed to process bioplastics and the more stable supply of biomass combined with the increasing cost of crude oil make bioplastics' prices more competitive with regular plastics.

Research and Development

- In the early 1950s, amylomaize (>50% amylose content corn) was successfully bred and commercial bioplastics applications started to be explored.

- In 2004, NEC developed a flame retardant plastic, polylactic acid, without using toxic chemicals such as halogens and phosphorus compounds.

Bioplastics Development Center - University of Massachusetts Lowell

A pen made with bioplastics (Polylactide, PLA)

- In 2005, Fujitsu became one of the first technology companies to make personal computer cases from bioplastics, which are featured in their FMV-BIBLO NB80K line. Later, the French company Ashelvea (also listed on EU Energy Star registered partners), launched its fully recyclable PC with biodegradable plastic case "Evolutis", reported in "People Inspiring Philips", a series of 3 mini-documentaries to inspire Philips employees with some examples from the civil society.

- In 2007 Braskem of Brazil announced it had developed a route to manufacture high-density polyethylene (HDPE) using ethylene derived from sugar cane.

- In 2008, a University of Warwick team created a soap-free emulsion polymerization process which makes colloid particles of polymer dispersed in water, and in a one step process adds nanometre sized silica-based particles to the mix. The newly developed technology might be most applicable to multi-layered biodegradable packaging, which could gain more robustness and water barrier characteristics through the addition of a nano-particle coating.

Testing Procedures

A bioplastic shampoo bottle made of PLA-blend bio-flex

Industrial Compostability – EN 13432, ASTM D6400

The EN 13432 industrial standard is arguably the most international in scope. This standard must be met in order to claim that a plastic product is compostable in the European marketplace. In summary, it requires biodegradation of 90% of the materials in a lab within 90 days. The ASTM 6400 standard is the regulatory framework for the United States and sets a less stringent threshold of 60% biodegradation within 180 days for non-homopolymers, and 90% biodegradation of homopolymers within industrial composting conditions (temperatures at or above 140F). Municipal compost facilities do not see above 130F.

Many starch-based plastics, PLA-based plastics and certain aliphatic-aromatic co-polyester compounds, such as succinates and adipates, have obtained these certificates. Additive-based bioplastics sold as photodegradable or Oxo Biodegradable do not comply with these standards in their current form.

Compostability – ASTM D6002

The ASTM D 6002 method for determining the compostability of a plastic defined the word *compostable* as follows:

that which is capable of undergoing biological decomposition in a compost site such that the material is not visually distinguishable and breaks down into carbon dioxide, water, inorganic compounds and biomass at a rate consistent with known compostable materials.

This definition drew much criticism because, contrary to the way the word is traditionally defined, it completely divorces the process of "composting" from the necessity of it leading to humus/compost as the end product. The only criterion this standard *does* describe is that a compostable plastic must look to be going away as fast as something else one has already established to be compostable under the *traditional* definition.

Withdrawal of ASTM D 6002

In January 2011, the ASTM withdrew standard ASTM D 6002, which had provided plastic manufacturers with the legal credibility to label a plastic as compostable. Its description is as follows:

This guide covered suggested criteria, procedures, and a general approach to establish the compostability of environmentally degradable plastics.

The ASTM has yet to replace this standard.

Biobased – ASTM D6866

The ASTM D6866 method has been developed to certify the biologically derived content of bioplastics. Cosmic rays colliding with the atmosphere mean that some of the carbon is the radioactive isotope carbon-14. CO_2 from the atmosphere is used by plants in photosynthesis, so new plant material will contain both carbon-14 and carbon-12. Under the right conditions, and over geological timescales, the remains of living organisms can be transformed into fossil fuels. After ~100,000 years all the carbon-14 present in the original organic material will have undergone radioactive decay leaving only carbon-12. A product made from biomass will have a relatively high level of carbon-14, while a product made from petrochemicals will have no carbon-14. The percentage of renewable carbon in a material (solid or liquid) can be measured with an accelerator mass spectrometer.

There is an important difference between biodegradability and biobased content. A bioplastic such as high-density polyethylene (HDPE) can be 100% biobased (i.e. contain 100% renewable carbon), yet be non-biodegradable. These bioplastics such as HDPE nonetheless play an important role in greenhouse gas abatement, particularly when they are combusted for energy production. The biobased component of these bioplastics is considered carbon-neutral since their origin is from biomass.

Anaerobic biodegradability – ASTM D5511-02 and ASTM D5526

The ASTM D5511-12 and ASTM D5526-12 are testing methods that comply with international standards such as the ISO DIS 15985 for the biodegradability of plastic.

Biodegradable Additives

Biodegradable additives are additives that enhance the biodegradation of polymers by allowing microorganisms to utilize the carbon within the polymer chain itself.

Biodegradable additives attract microorganisms to the polymer through quorum sensing after biofilm creation on the plastic product. Additives are generally in masterbatch formation that use carrier resins such as polyethylene, polypropylene, polystyrene or polyethylene terephthalate.

Testing Methods of Biodegradable Additives

ASTM D5511-12 testing is for the "Anerobic Biodegradation of Plastic Materials in a High Solids Environment Under High-Solids Anaerobic-Digestion Conditions"

ASTM D5526-12 testing is for the "Standard Test Method for Determining Anaerobic Biodegradation of Plastic Materials Under Accelerated Landfill Conditions"

ASTM D5210-07 testing is for the "Standard Test Method for Determining the Anaerobic Biodegradation of Plastic Materials in the Presence of Municipal Sewage Sludge"

Laboratories Performing ASTM Testing Methods

- Eden Research Labs
- Respirtek
- NE Laboratories
- NSF

Biodegradation Process of Biodegradable Additives

A simple chemical equation of the process is:

$$C_6H_{12}O_6 \rightarrow 3CO_2 + 3CH_4$$

Interpretation of this process is as follows - In most cases, plastic is made up of hydrophobic polymers. Chains must be broken down into constituent parts for the energy potential to be used by microorganisms. These constituent parts, or monomers, are readily available to other bacteria. The process of breaking these chains and dissolving the smaller molecules into solution is called hydrolysis. Therefore, hydrolysis of these high-molecular-weight polymeric components is the necessary first step in anaerobic biodegradation. Through hydrolysis, the complex organic molecules are broken down into simple sugars, amino acids, and fatty acids.

Acetate and hydrogen produced in the first stages can be used directly by methanogens. Other molecules, such as volatile fatty acids (VFAs) with a chain length greater than that of acetate must first be catabolised into compounds that can be directly used by methanogens.

The biological process of acidogenesis results in further breakdown of the remaining components by acidogenic (fermentative) bacteria. Here, VFAs are created, along with ammonia, carbon dioxide, and hydrogen sulfide, as well as other byproducts. The process of acidogenesis is similar to the way milk sours.

The third stage of anaerobic digestion is acetogenesis. Simple molecules created through the acidogenesis phase are further digested by Acetogens to produce largely acetic acid, as well as carbon dioxide and hydrogen.

The terminal stage of anaerobic biodegradation is the biological process of methanogenesis. Here, methanogens use the intermediate products of the preceding stages and convert them into methane, carbon dioxide, and water. These components make up the majority of the biogas emitted. Methanogenesis is sensitive to both high and low pHs and occurs between pH 6.5 and pH 8. The remaining, indigestible material the microbes cannot use and any dead bacterial remains constitute the digestate.

Biodegradable Additive Manufacturers

- Bio-Tec Environmental, LLC
- EcoLogic LLC
- EcoSafe Plastic
- BioSphere Plastic

- ENSO Plastics
- Hybrid Green

Polyhydroxybutyrate

Polyhydroxybutyrate (PHB) is a polyhydroxyalkanoate (PHA), a polymer belonging to the polyesters class that are of interest as bio-derived and biodegradable plastics. The poly-3-hydroxybutyrate (P3HB) form of PHB is probably the most common type of polyhydroxyalkanoate, but other polymers of this class are produced by a variety of organisms: these include poly-4-hydroxybutyrate (P4HB), polyhydroxyvalerate (PHV), polyhydroxyhexanoate (PHH), polyhydroxyoctanoate (PHO) and their copolymers.

Biosynthesis

PHB is produced by microorganisms (such as *Ralstonia eutrophus, Methylobacterium rhodesianum* or *Bacillus megaterium*) apparently in response to conditions of physiological stress; mainly conditions in which nutrients are limited. The polymer is primarily a product of carbon assimilation (from glucose or starch) and is employed by microorganisms as a form of energy storage molecule to be metabolized when other common energy sources are not available.

Microbial biosynthesis of PHB starts with the condensation of two molecules of acetyl-CoA to give acetoacetyl-CoA which is subsequently reduced to hydroxybutyryl-CoA. This latter compound is then used as a monomer to polymerize PHB. PHAs granules are then recovered by disrupting the cells.

Structure of poly-(R)-3-hydroxybutyrate (P3HB), a polyhydroxyalkanoate

Chemical structures of P3HB, PHV and their copolymer PHBV

Thermoplastic Polymer

Most commercial plastics are synthetic polymers derived from petrochemicals. They tend to resist biodegradation. PHB-derived plastics are attractive because they are compostable and derived from renewables and are bio-degradable.

ICI had developed the material to pilot plant stage in the 1980s, but interest faded when it became clear that the cost of material was too high, and its properties could not match those of polypropylene.

In 1996 Monsanto (who sold PHB as a copolymer with PHV under the trade name Biopol) bought all patents for making the polymer from ICI/Zeneca. However, Monsanto's rights to Biopol were sold to the American company Metabolix in 2001 and Monsanto's fermenters producing PHB from bacteria were closed down at the start of 2004. Monsanto began to focus on producing PHB from plants instead of bacteria. But now with so much media attention on GM crops, there has been little news of Monsanto's plans for PHB.

In June 2005, a US company, Metabolix, received the Presidential Green Chemistry Challenge Award (small business category) for their development and commercialisation of a cost-effective method for manufacturing PHAs in general, including PHB.

Biopol is currently used in the medical industry for internal suture. It is nontoxic and biodegradable, so it does not have to be removed after recovery.

History

Polyhydroxybutyrate was first isolated and characterized in 1925 by French microbiologist Maurice Lemoigne.

Biodegradation

Firmicutes and proteobacteria can degrade PHB. Bacillus, Pseudomonas and Streptomyces species can degrade PHB. Pseudomonas lemoigne, Comamonas sp. Acidovorax faecalis, Aspergillus fumigatus and Variovorax paradoxus are soil microbes capable of degradation. Alcaligenes faecalis, Pseudomonas, and Illyobacter delafieldi, are obtained from anaerobic sludge. Comamonas testosterone and Pseudomonas stutzeri were obtained from sea water. Few of these are capable of degrading at higher temperatures; notably excepting thermophilic Streptomyces sp. and a thermophilic strain of Aspergillus sp

Polylactic Acid

The name "polylactic acid" does not comply with IUPAC standard nomenclature, and is potentially ambiguous or confusing, because PLA is not a polyacid (polyelectrolyte), but rather a polyester.

Polylactic acid

Poly(lactic acid) or polylactic acid or polylactide (PLA) is a biodegradable thermoplastic aliphatic polyester derived from renewable resources, such as corn starch (in the United States and Canada), tapioca roots, chips or starch (mostly in Asia), or sugarcane (in the rest of the world). In 2010, PLA had the second highest consumption volume of any bioplastic of the world.

Production

Producers have several industrial routes to usable (i.e. high molecular weight) PLA. Two main monomers are used: lactic acid, and the cyclic di-ester, lactide. The most common route to PLA is the ring-opening polymerization of lactide with various metal catalysts (typically tin octoate) in solution, in the melt, or as a suspension. The metal-catalyzed reaction tends to cause racemization of the PLA, reducing its stereoregularity compared to the starting material (usually corn starch).

Another route to PLA is the direct condensation of lactic acid monomers. This process needs to be carried out at less than 200 °C; above that temperature, the entropically favored lactide monomer is generated. This reaction generates one equivalent of water for every condensation (esterification) step, and that is undesirable because water causes chain-transfer leading to low molecular weight material. The direct condensation is thus performed in a stepwise fashion, where lactic acid is first oligomerized to PLA oligomers. Thereafter, polycondensation is done in the melt or as a solution, where short oligomeric units are combined to give a high molecular weight polymer strand. Water removal by application of a vacuum or by azeotropic distillation is crucial to favor polycondensation over transesterification. Molecular weights of 130 kDa can be obtained this way. Even higher molecular weights can be attained by carefully crystallizing the crude polymer from the melt. Carboxylic acid and alcohol end groups are thus concentrated in the amorphous region of the solid polymer, and so they can react. Molecular weights of 128–152 kDa are obtainable thus.

Polymerization of a racemic mixture of L- and D-lactides usually leads to the synthesis of poly-DL-lactide (PDLLA), which is amorphous. Use of stereospecific catalysts can lead to heterotactic PLA which has been found to show crystallinity. The degree of crystallinity, and hence many important properties, is largely controlled by the ratio of D to L enantiomers used, and to a lesser extent on the type of catalyst used. Apart from lactic acid and lactide, lactic acid O-carboxyanhydride ("lac-OCA"), a five-membered cyclic compound has been used academically as well. This compound is more reactive than lactide, because its polymerization is driven by the loss of one equivalent of carbon dioxide per equivalent of lactic acid. Water is not a co-product.

The direct biosynthesis of PLA similar to the poly(hydroxyalkanoate)s has been reported as well.

Manufacturers

As of June 2010, NatureWorks was the primary producer of PLA (bioplastic) in the United States.

Other companies involved in PLA manufacturing are Evonik Industries (Germany), Corbion PURAC Biomaterials (The Netherlands) who have announced a new 75,000 ton PLA plant in Thailand by 2018, and several Chinese manufacturers. The primary producer of PDLLA is Evonik Industries and Corbion PURAC. Evonik Industries is a specialty chemical company that is industry leading in approximately 80% of the markets they participate. The Resomer brand of PDLLA is produced in the Health and Nutrition business segment. Corbion PURAC is a listed company in the Netherlands, and operating plants worldwide, and the only produce of PDLA, produced from the D-isomer of lactid acid. Galactic and Total Petrochemicals operate a joint venture, Futerro, which is developing a second generation polylactic acid product. This project includes the building of a PLA pilot plant in Belgium capable of producing 1,500 tonnes/year.

Chemical and Physical Properties

Due to the chiral nature of lactic acid, several distinct forms of polylactide exist: poly-L-lactide (PLLA) is the product resulting from polymerization of L,L-lactide (also known as L-lactide). PLLA has a crystallinity of around 37%, a glass transition temperature 60–65 °C, a melting temperature 173–178 °C and a tensile modulus 2.7–16 GPa. Heat-resistant PLA can withstand temperatures of 110 °C. PLA is soluble in chlorinated solvents, hot benzene, tetrahydrofuran, and dioxane.

Polylactic acid can be processed like most thermoplastics into fiber (for example, using conventional melt spinning processes) and film. PLA has similar mechanical properties to PETE polymer, but has a significantly lower maximum continuous use temperature. The tensile strength for 3-D printed PLA was previously determined. It was found to range widely depending on printing conditions, which were obtained using RepRap 3-D printers. Results of a recent study gave a printed tensile strength of around 50 MPa and show that the act of 3-D printing PLA affects its properties—they showed a strong relationship between tensile strength and percent crystallinity of a 3-D printed sample and a strong relationship between percent crystallinity and the extruder temperature.

The melting temperature of PLLA can be increased by 40–50 °C and its heat deflection temperature can be increased from approximately 60 °C to up to 190 °C by physically blending the polymer with PDLA (poly-D-lactide). PDLA and PLLA form a highly regular stereocomplex with increased crystallinity. The temperature stability is maximised when a 1:1 blend is used, but even at lower concentrations of 3–10% of PDLA, there is still a substantial improvement. In the latter case, PDLA acts as a nucleating agent, thereby increasing the crystallization rate. Biodegradation of PDLA is slower than for PLA due to the higher crystallinity of PDLA.

There is also poly(L-lactide-*co*-D,L-lactide) (PLDLLA) – used as PLDLLA/TCP scaffolds for bone engineering.

Applications

PLA can be processed by extrusion such as 3d printing, injection molding, film and sheet casting, and spinning, providing access to a wide range of materials.

Biodegradable PLA cups in use at a restaurant

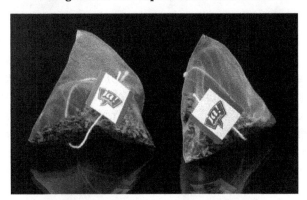

Tea bags made of PLA. Peppermint tea is enclosed.

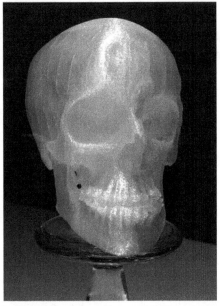

3D Printed Human skull with data from Computed Tomography. Transparent PLA.

PLA is used as a feedstock material in desktop fused filament fabrication-based 3D printers (e.g. RepRap). PLA printed solids can be encased in plaster-like moulding materials, then burned out in a furnace, so that the resulting void can be filled with molten metal. This is known as "lost PLA casting", a type of investment casting.

Being able to degrade into innocuous lactic acid, PLA is used as medical implants in the form of anchors, screws, plates, pins, rods, and as a mesh. Depending on the exact type used, it breaks down inside the body within 6 months to 2 years. This gradual degradation is desirable for a support structure, because it gradually transfers the load to the body (e.g. the bone) as that area heals. The strength characteristics of PLA and PLLA implants are well documented.

PLA can also be used as a decomposable packaging material, either cast, injection-molded, or spun. Cups and bags have been made from this material. In the form of a film, it shrinks upon heating, allowing it to be used in shrink tunnels. It is useful for producing loose-fill packaging, compost bags, food packaging, and disposable tableware. In the form of fibers and nonwoven fabrics, PLA also has many potential uses, for example as upholstery, disposable garments, awnings, feminine hygiene products, and diapers.

Racemic and regular PLLA has a low glass transition temperature, which is undesirable. A stereocomplex of PDLA and PLLA has a higher glass transition temperatures, lending it more mechanical strength. It has a wide range of applications, such as woven shirts (ironability), microwavable trays, hot-fill applications and even engineering plastics (in this case, the stereocomplex is blended with a rubber-like polymer such as ABS). Such blends also have good form stability and visual transparency, making them useful for low-end packaging applications. Pure poly-L-lactic acid (PLLA), on the other hand, is the main ingredient in Sculptra, a long-lasting facial volume enhancer, primarily used for lipoatrophy of cheeks. Progress in biotechnology has resulted in the development of commercial production of the D enantiomer form, something that was not possible until recently.

Recycling

Currently, the SPI resin identification code 7 ("others") is applicable for PLA. In Belgium, Galactic started the first pilot unit to chemically recycle PLA (Loopla). Unlike mechanical recycling, waste material can hold various contaminants. Polylactic acid can be recycled to monomer by thermal depolymerization or hydrolysis. When purified, the monomer can be used for the manufacturing of virgin PLA with no loss of original properties (cradle-to-cradle recycling).

PLA has SPI resin ID code 7

Degradation

Amycolatopsis and *Saccharotrix* are able to degrade PLA. A purified protease from *Amycolatopsis* sp., PLA depolymerase, can also degrade PLA. Enzymes such as pronase and most effectively proteinase K from *Tritirachium album* degrade PLA.

Pure PLLA foams undergo selective hydrolysis when placed in an environment of Dulbecco's modified Eagle's medium (DMEM) supplemented with fetal bovine serum (FBS) (a solution mimicking body fluid). After 30 days of submersion in DMEM+FBS, a PLLA scaffold lost about 20% of its weight.

Polycaprolactone

This polymer is often used as an additive for resins to improve their processing characteristics and their end use properties (e.g., impact resistance). Being compatible with a range of other materials, PCL can be mixed with starch to lower its cost and increase biodegradability or it can be added as a polymeric plasticizer to PVC.

Polycaprolactone

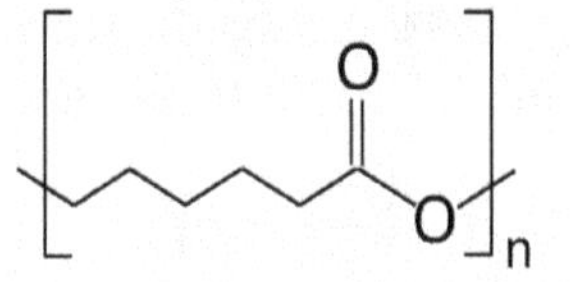

Polycaprolactone (PCL) is a biodegradable polyester with a low melting point of around 60 °C and a glass transition temperature of about −60 °C. The most common use of polycaprolactone is in the manufacture of speciality polyurethanes. Polycaprolactones impart good water, oil, solvent and chlorine resistance to the polyurethane produced.

Polycaprolactone is also used for splinting, modeling, and as a feedstock for prototyping systems such as Fused Filament Fabrication 3D Printers.

Synthesis

PCL is prepared by ring opening polymerization of ε-caprolactone using a catalyst such as stannous octoate. Recently a wide range of catalysts for the ring opening polymerization of caprolactone have been reviewed.

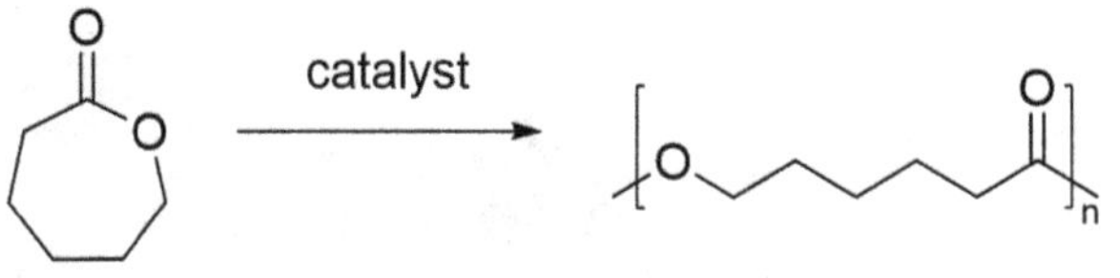

Biomedical applications

PCL is degraded by hydrolysis of its ester linkages in physiological conditions (such as in the human body) and has therefore received a great deal of attention for use as an implantable biomaterial. In particular it is especially interesting for the preparation of long term implantable devices, owing to its degradation which is even slower than that of polylactide.

PCL has been approved by the Food and Drug Administration (FDA) in specific applications used in the human body as (for example) a drug delivery device, suture (sold under the brand name Monocryl or generically), or adhesion barrier. It is being investigated as a scaffold for tissue repair via tissue engineering, GBR membrane. It has been used as the hydrophobic block of amphiphilic synthetic block copolymers used to form the vesicle membrane of polymersomes.

It is also used in housing applications.

A variety of drugs have been encapsulated within PCL beads for controlled release and targeted drug delivery.

The major impurities in the medical grade are toluene (<890 ppm, usually about 100 ppm) and tin (<200 ppm).

In dentistry (as composite named Resilon), it is used as a component of "night guards" (dental splints) and in root canal filling. It performs like gutta-percha, has similar handling properties, and for retreatment purposes may be softened with heat, or dissolved with solvents like chloroform. Similar to gutta-percha, there are master cones in all ISO sizes and accessory cones in different sizes and taper available. The major difference between the polycaprolactone-based root canal filling material (Resilon and Real Seal) and gutta-percha is that the PCL-based material is biodegradable but gutta-percha is not. There is lack of consensus in the expert dental community as to whether a biodegradable root canal filling material, such as Resilon or Real Seal is desirable.

Hobbyist and Prototyping

PCL also has many applications in the hobbyist market where it is known as Polymorph, Shapelock, ReMoldables, Plastdude or TechTack. It has physical properties of a very tough, nylon-like plastic that softens to a putty-like consistency at only 60 °C, easily achieved by immersing in hot water. PCL's specific heat and conductivity are low enough that it is not hard to handle by hand at this temperature. This makes it ideal for small-scale modeling, part fabrication, repair of plastic objects, and rapid prototyping where heat resistance is not needed. Though softened PCL readily sticks to many other plastics when at higher temperature, if the surface is cooled, the stickiness can be minimized while still leaving the mass pliable.

Home-made bicycle light mounting, made from PCL

Biodegradation

Firmicutes and proteobacteria can degrade PCL. Penicillium sp. strain 26-1 can degrade high density PCL; though not as quickly as thermotolerant Aspergillus sp. strain ST-01. Species of clostridium can degrade PCL under anaerobic conditions.

Polyvinyl Alcohol

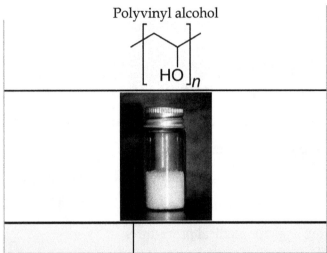

Polyvinyl alcohol

Poly(vinyl alcohol) (PVOH, PVA, or PVAl) is a water-soluble synthetic polymer. It has the idealized formula $[CH_2CH(OH)]_n$. It is used in papermaking, textiles, and a variety of coatings. It is white (colourless) and odorless. It is sometimes supplied as beads or as solutions in water.

Uses

- Polyvinyl acetals: Polyvinyl acetals are prepared by reacting aldehydes with polyvinyl alcohol. Polyvinyl butyral (PVB) and polyvinyl formal (PVF) are examples of this family of polymers. They are prepared from polyvinyl alcohol by reaction with butyraldehyde and formaldehyde, respectively. Preparation of polyvinyl butyral is the largest use for polyvinyl alcohol in the U.S. and Western Europe.

Polyvinyl alcohol is used as an emulsion polymerization aid, as protective colloid, to make polyvinyl acetate dispersions. This is the largest market application in China. In Japan its major use is vinylon fiber production.

Some other uses of polyvinyl alcohol include:

- Paper adhesive with boric acid in spiral tube winding and solid board production

- Thickener, modifier, in polyvinyl acetate glues

- Textile sizing agent

- Paper coatings, release liner

- As a water-soluble film useful for packaging. An example is the envelope containing laundry detergent in "liqui-tabs".

- Feminine hygiene and adult incontinence products as a biodegradable plastic backing sheet.

- Carbon dioxide barrier in polyethylene terephthalate (PET) bottles

- As a film used in the water transfer printing process

- As a form release because materials such as epoxy do not stick to it

- Movie practical effect and children's play putty or slime when combined with borax

- Used in eye drops (such as artificial tears to treat dry eyes) and hard contact lens solution as a lubricant

- PVA fiber, as reinforcement in concrete

- Raw material to polyvinyl nitrate (PVN) an ester of nitric acid and polyvinyl alcohol.

- As a surfactant for the formation of polymer encapsulated nanobeads

- Used in protective chemical-resistant gloves

- Used as a fixative for specimen collection, especially stool samples

- When doped with iodine, PVA can be used to polarize light

- As an embolization agent in medical procedures

- Carotid phantoms for use as synthetic vessels in Doppler flow testing

- Used in 3D printing as support structure that can then be dissolved away.

Fishing

PVA is widely used in freshwater sport fishing. Small bags made from PVA are filled with dry or oil based bait and attached to the hook, or the baited hook is placed inside the bag and cast into the water. When the bag lands on the lake or river bottom it breaks down, leaving the hook bait surrounded by ground bait, pellets etc. This method helps attract fish to the hook bait.

Anglers also use string made of PVA for the purpose of making temporary attachments. For example, holding a length of line in a coil, that might otherwise tangle while the cast is made.

Preparation

Unlike most vinyl polymers, PVA is not prepared by polymerization of the corresponding monomer. The monomer, vinyl alcohol, is unstable with respect to acetaldehyde. PVA instead is prepared by first polymerizing vinyl acetate, and the resulting polyvinylacetate is converted to the PVA. Other precursor polymers are sometimes used, with formate, chloroacetate groups instead of acetate. The conversion of the polyesters is usually conducted by base-catalysed transesterification with ethanol:

$$[CH_2CH(OAc)]_n + C_2H_5OH \rightarrow [CH_2CH(OH)]_n + C_2H_5OAc$$

The properties of the polymer depend on the amount of residual ester groups.

Worldwide consumption of polyvinyl alcohol was over one million metric tons in 2006. Larger producers include Kuraray (Japan and Europe) and Sekisui Specialty Chemicals (USA) but mainland China has installed a number of very large production facilities in the past decade and currently accounts for 45% of world capacity. The North Korean-manufacture fiber Vinalon is produced from polyvinyl alcohol. Despite its inferior properties as a clothing fiber, it is produced for self-sufficiency reasons, because no oil is required to produce it.

Structure and Properties

PVA is an atactic material that exhibits crystallinity. In terms of microstructure, it is composed mainly of 1,3-diol linkages [-CH$_2$-CH(OH)-CH$_2$-CH(OH)-] but a few percent of 1,2-diols [-CH$_2$-CH(OH)-CH(OH)-CH$_2$-] occur, depending on the conditions for the polymerization of the vinyl ester precursor.

Polyvinyl alcohol has excellent film forming, emulsifying and adhesive properties. It is also resistant to oil, grease and solvents. It has high tensile strength and flexibility, as well as high oxygen and aroma barrier properties. However these properties are dependent on humidity, in other words, with higher humidity more water is absorbed. The water, which acts as a plasticiser, will then reduce its tensile strength, but increase its elongation and tear strength.

PVA has a melting point of 230 °C and 180–190 °C (356-374 degrees Fahrenheit) for the fully hydrolysed and partially hydrolysed grades, respectively. It decomposes rapidly above 200 °C as it can undergo pyrolysis at high temperatures.

PVA is close to incompressible. The Poisson's ratio is between 0.42 and 0.48.

Tradenames of Polyvinyl Alcohol

Kuraray Poval, Mowiol, Celvol, Polyviol

Safety

PVA is nontoxic. It biodegrades slowly, and solutions containing up to 5% PVA are nontoxic to fish.

Polypropylene Carbonate

Polypropylene carbonate

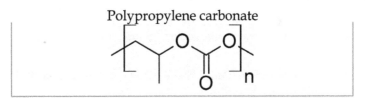

Polypropylene carbonate (PPC), a copolymer of carbon dioxide and propylene oxide, is a thermoplastic material. Catalysts like zinc glutarate are used in polymerization.

Properties

Polypropylene carbonate is soluble in polar solvents like lower ketones, ethyl acetate, dichloromethane and chlorinated hydrocarbons and insoluble in solvents like alcohols, water, and aliphatic hydrocarbons. It also forms stable emulsions in water. PPC allows the diffusion of gases like oxygen through it. Having a glass temperature (T_g) between 25 to 45 °C, PPC binders are amorphous. The glass temperature of PPC is slightly greater than polyethylene carbonate (PEC).

Its refractive index is 1.46 while its dielectric constant is 3.

Applications

Polypropylene carbonate is used to increase the toughness of some epoxy resins. It is used as a sacrificial binder in the ceramic industry, which decomposes and evaporates during sintering. It has a low sodium content which makes it suitable for the preparation of electroceramics like dielectric materials and piezoelectric ceramics.

Composites of polypropylene carbonate with starch (PPC/starch) are used as biodegradable plastics.

One of the largest manufacturers of polypropylene carbonate is Empower Materials, located in New Castle, DE, USA.

References

- Rafael Auras; Loong-Tak Lim; Susan E. M. Selke; Hideto Tsuji (eds.). Poly(Lactic Acid): Synthesis, Structures, Properties, Processing, and Applications. doi:10.1002/9780470649848. ISBN 9780470293669.

- Nazre, A.; Lin, S. (1994). Harvey, J. Paul; Games, Robert F., eds. Theoretical Strength Comparison of Bioabsorbable (PLLA) Plates and Conventional Stainless Steel and Titanium Plates Used in Internal Fracture Fixation. p. 53. ISBN 0-8031-1897-X.

- Labet, Marianne; Thielemans, Wim (2009). "Synthesis of polycaprolactone: a review". Chemical Society Reviews. 38 (12): 3484–3504. doi:10.1039/B820162P. PMID 20449064.

- Supercilii, Corrugator. "DIY Material Guide: Polymorph Plastic (a thermal plastic with low melting point)". Instructables. Autodesk. Retrieved 20 August 2015.

- Khwaldia, Khaoula; Elmira Arab-Tehrany; Stephane Desobry (2010). "Biopolymer Coatings on Paper Packaging Materials". Comprehensive Reviews in Food Science and Food Safety. 9 (1): 82–91. doi:10.1111/j.1541-4337.2009.00095.x. Retrieved 9 Mar 2015.

- Fortman, David J.; Jacob P. Brutman; Christopher J. Cramer; Marc A. Hillmyer; William R. Dichtel (2015). "Mechanically Activated, Catalyst-Free Polyhydroxyurethane Vitrimers". Journal of the American Chemical Society. 137: 14019–14022. doi:10.1021/jacs.5b08084.

- Guo, Shuang-Zhuang; Yang, Xuelu; Heuzey, Marie-Claude; Therriault, Daniel (2015). "3D printing of a multifunctional nanocomposite helical liquid sensor". Nanoscale. 7 (15): 6451. doi:10.1039/C5NR00278H. PMID 25793923.

- Ghosh, Sudhipto. "European Parliament Committee Vote for 100% Biodegradable Plastic Bags." Modern Plastics and Polymers. Network 18, 19 Mar. 2014.

- Nohra, Bassam; Laure Candy; Jean-Francois Blanco; Celine Guerin; Yann Raoul; Zephirin Mouloungui (2013). "From Petrochemical Polyurethanes to Biobased Polyhydroxyurethanes". Macromolecules. 46: 3771–3792. doi:10.1021/ma400197c.

- "Bioplastic creates Nitrous Oxide" (PDF). American Carbon Registry. Archived from the original (PDF) on December 2, 2012. Retrieved 2013-01-10.

- Bose, S., Vahabzadeh, S. and Bandyopadhyay, A., 2013. Bone tissue engineering using 3D printing. Materials Today, 16(12), pp.496-504.

- William Harris. "How long does it take for plastics to biodegrade?". How Stuff Works. Retrieved 2013-05-09.

- "Terminology for biorelated polymers and applications (IUPAC Recommendations 2012)" (PDF). Pure and Applied Chemistry. 84 (2): 377–410. 2012. doi:10.1351/PAC-REC-10-12-04.

- Chen, G.; Patel, M. (2012). "Plastics derived from biological sources: Present and future: P technical and environmental review". Chemical Reviews. 112 (4): 2082–2099. doi:10.1021/cr200162d.

- Gerrit. A. Luinstra, Endres Borchardt. "Material Properties of Poly(Propylene Carbonates)" (PDF). Retrieved July 10, 2012.

- ASTM D6002 - 96(2002)e1 Standard Guide for Assessing the Compostability of Environmentally Degradable Plastics (Withdrawn 2011)

4

Essential Concepts of Biodegradable Waste Management

The major concepts of biodegradable waste management are discussed in this chapter. Biodegradable waste includes organic matter that can be broken down into simple organic molecules, carbon dioxide or water by microorganism and other living things. The significant concepts dealt here are waste management, bioconversion of biomass to mixed alcohol fuels, biofuel etc.

Waste Management

Waste management is all the activities and actions required to manage waste from its inception to its final disposal. This includes amongst other things, collection, transport, treatment and disposal of waste together with monitoring and regulation. It also encompasses the legal and regulatory framework that relates to waste management encompassing guidance on recycling etc.

Waste management in Kathmandu, Nepal

The term usually relates to all kinds of waste, whether generated during the extraction of raw materials, the processing of raw materials into intermediate and final products, the consumption of final products, or other human activities, including municipal (residential, institutional, commercial), agricultural, and social (health care, household hazardous wastes, sewage sludge). Waste management is intended to reduce adverse effects of waste on health, the environment or aesthetics.

Waste management practices are not uniform among countries (developed and developing nations); regions (urban and rural area), and sectors (residential and industrial).

Waste management in Stockholm, Sweden

Central Principles of Waste Management

There are a number of concepts about waste management which vary in their usage between countries or regions. Some of the most general, widely used concepts include:

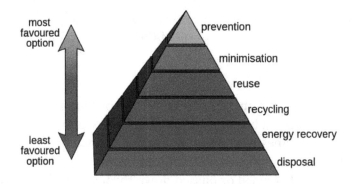

Diagram of the waste hierarchy

Waste Hierarchy

The waste hierarchy refers to the "3 Rs" reduce, reuse and recycle, which classify waste management strategies according to their desirability in terms of waste minimisation. The waste hierarchy remains the cornerstone of most waste minimisation strategies. The aim of the waste hierarchy is to extract the maximum practical benefits from products and to generate the minimum amount of waste. The waste hierarchy is represented as a pyramid because the basic premise is for policy to take action first and prevent the generation of waste. The next step or pre-ferred action is to reduce the generation of waste i.e. by re-use. The next is recycling which would include composting. Following this step is material recovery and waste-to-energy. Energy can be recovered from processes i.e. landfill and combustion, at this level of the hierarchy. The final action is disposal, in landfills or through incineration without energy recovery. This last step is the final resort for waste which has not been prevented, diverted or recovered. The waste hierarchy represents the progression of a product or material through the sequential stages of the pyramid of waste management. The hierarchy represents the latter parts of the life-cycle for each product.

Life-Cycle of A Product

The life-cycle begins with design, then proceeds through manufacture, distribution, use and then follows through the waste hierarchy's stages of reuse, recovery, recycling and disposal. Each of the above stages of the life-cycle offers opportunities for policy intervention, to rethink the need for the product, to redesign to minimize waste potential, to extend its use. The key behind the life-cycle of a product is to optimize the use of the world's limited resources by avoiding the unnecessary generation of waste.

Resource Efficiency

Resource efficiency reflects the understanding that current, global, economic growth and development can not be sustained with the current production and consumption patterns. Globally, we are extracting more resources to produce goods than the planet can replenish. Resource efficiency is the reduction of the environmental impact from the production and consumption of these goods, from final raw material extraction to last use and disposal. This process of resource efficiency can address sustainability.

Polluter Pays Principle

The Polluter pays principle is a principle where the polluting party pays for the impact caused to the environment. With respect to waste management, this generally refers to the requirement for a waste generator to pay for appropriate disposal of the unrecoverable material.

History

Throughout most of history, the amount of waste generated by humans was insignificant due to low population density and low societal levels of the exploitation of natural resources. Common waste produced during pre-modern times was mainly ashes and human biodegradable waste, and these were released back into the ground locally, with minimum environmental impact. Tools made out of wood or metal were generally reused or passed down through the generations.

However, some civilizations do seem to have been more profligate in their waste output than others. In particular, the Maya of Central America had a fixed monthly ritual, in which the people of the village would gather together and burn their rubbish in large dumps.

Modern Era

Sir Edwin Chadwick's 1842 report *The Sanitary Condition of the Labouring Population* was influential in securing the passage of the first legislation aimed at waste clearance and disposal.

Following the onset of industrialisation and the sustained urban growth of large population centres in England, the buildup of waste in the cities caused a rapid deterioration in levels of sanitation and the general quality of urban life. The streets became choked with filth due to the lack of waste clearance regulations. Calls for the establishment of a municipal authority with waste removal powers occurred as early as 1751, when Corbyn Morris in London proposed that "...as the preservation of the health of the people is of great importance, it is proposed that the cleaning of this city, should be put under one uniform public management, and all the filth be...conveyed by the Thames to proper distance in the country".

However, it was not until the mid-19th century, spurred by increasingly devastating cholera outbreaks and the emergence of a public health debate that the first legislation on the issue emerged. Highly influential in this new focus was the report *The Sanitary Condition of the Labouring Population* in 1842 of the social reformer, Edwin Chadwick, in which he argued for the importance of adequate waste removal and management facilities to improve the health and wellbeing of the city's population.

In the UK, the Nuisance Removal and Disease Prevention Act of 1846 began what was to be a steadily evolving process of the provision of regulated waste management in London. The Metropolitan Board of Works was the first city-wide authority that centralized sanitation regulation for the rapidly expanding city and the Public Health Act 1875 made it compulsory for every household to deposit their weekly waste in "moveable receptacles: for disposal—the first concept for a dust-bin.

Manlove, Alliott & Co. Ltd. 1894 destructor furnace. The use of incinerators for waste disposal became popular in the late 19th century.

The dramatic increase in waste for disposal led to the creation of the first incineration plants, or, as they were then called, "destructors". In 1874, the first incinerator was built in Nottingham by Manlove, Alliott & Co. Ltd. to the design of Albert Fryer. However, these were met with opposition on account of the large amounts of ash they produced and which wafted over the neighbouring areas.

Similar municipal systems of waste disposal sprung up at the turn of the 20th century in other large cities of Europe and North America. In 1895, New York City became the first U.S. city with public-sector garbage management.

Early garbage removal trucks were simply open bodied dump trucks pulled by a team of horses. They became motorized in the early part of the 20th century and the first close body trucks to eliminate odours with a dumping lever mechanism were introduced in the 1920s in Britain. These were soon equipped with 'hopper mechanisms' where the scooper was loaded at floor level and then hoisted mechanically to deposit the waste in the truck. The Garwood Load Packer was the first truck in 1938, to incorporate a hydraulic compactor.

Waste Handling and Transport

Waste collection methods vary widely among different countries and regions. Domestic waste collection services are often provided by local government authorities, or by private companies for industrial and commercial waste. Some areas, especially those in less developed countries, do not have formal waste-collection systems.

Molded plastic, wheeled waste bin in Berkshire, England

Waste Handling Practices

Curbside collection is the most common method of disposal in most European countries, Canada, New Zealand and many other parts of the developed world in which waste is collected at regular intervals by specialised trucks. This is often associated with curb-side waste segregation. In rural areas waste may need to be taken to a transfer station. Waste collected is then transported to an appropriate disposal facility. In some areas, vacuum collection is used in which waste is transported from the home or commercial premises by vacuum along small bore tubes. Systems are in use in Europe and North America.

Pyrolysis is used to dispose of some wastes including tires, a process that can produce recovered fuels, steel and heat. In some cases tires can provide the feedstock for cement manufacture. Such systems are used in USA, California, Australia, Greece, Mexico, the United Kingdom and in Israel. The RESEM pyrolysis plant that has been operational at Texas USA since December 2011, and processes up to 60 tons per day. In some jurisdictions unsegregated waste is collected at the curb-side or from waste transfer stations and then sorted into recyclables and unusable waste. Such systems are capable of sorting large volumes of solid waste, salvaging recyclables, and turning the rest into bio-gas and soil conditioner. In San Francisco, the local government established its Mandatory Recycling and Composting Ordinance in support of its goal of zero waste by 2020, requiring everyone in the city to keep recyclables and compostables out of the landfill. The three streams are collected with the curbside "Fantastic 3" bin system - blue for recyclables, green for compostables, and black for landfill-bound materials - provided to residents and businesses and serviced by San Francisco's sole refuse hauler, Recology. The City's "Pay-As-You-Throw" system charges customers by the volume of landfill-bound materials, which provides a financial incentive to separate recyclables and compostables from other discards. The City's Department of the Environment's Zero Waste Program has led the City to achieve 80% diversion, the highest diversion rate in North America.

Financial Models

In most developed countries, domestic waste disposal is funded from a national or local tax which may be related to income, or notional house value. Commercial and industrial waste disposal is typically charged for as a commercial service, often as an integrated charge which includes disposal costs. This practice may encourage disposal contractors to opt for the cheapest disposal option such as landfill rather than the environmentally best solution such as re-use and recycling. In some areas such as Taipei, the city government charges its households and industries for the volume of rubbish they produce. Waste will only be collected by the city council if waste is disposed in government issued rubbish bags. This policy has successfully reduced the amount of waste the city produces and increased the recycling rate.

Disposal Solutions

Landfill

A landfill compaction vehicle in action.

Spittelau incineration plant in Vienna

Incineration

Incineration is a disposal method in which solid organic wastes are subjected to combustion so as to convert them into residue and gaseous products. This method is useful for disposal of residue of both solid waste management and solid residue from waste water management. This process reduces the volumes of solid waste to 20 to 30 percent of the original volume. Incineration and other high temperature waste treatment systems are sometimes described as "thermal treatment". Incinerators convert waste materials into heat, gas, steam, and ash.

Incineration is carried out both on a small scale by individuals and on a large scale by industry. It is used to dispose of solid, liquid and gaseous waste. It is recognized as a practical method of disposing of certain hazardous waste materials (such as biological medical waste). Incineration is a controversial method of waste disposal, due to issues such as emission of gaseous pollutants.

Incineration is common in countries such as Japan where land is more scarce, as these facilities generally do not require as much area as landfills. Waste-to-energy (WtE) or energy-from-waste (EfW) are broad terms for facilities that burn waste in a furnace or boiler to generate heat, steam or electricity. Combustion in an incinerator is not always perfect and there have been concerns about pollutants in gaseous emissions from incinerator stacks. Particular concern has focused on some very persistent organic compounds such as dioxins, furans, and PAHs, which may be created and which may have serious environmental consequences.

Recycling

Recycling is a resource recovery practice that refers to the collection and reuse of waste materials such as empty beverage containers. The materials from which the items are made can be reprocessed into new products. Material for recycling may be collected separately from general waste using dedicated bins and collection vehicles, a procedure called kerbside collection. In some communities, the owner of the waste is required to separate the materials into different bins (e.g. for paper, plastics, metals) prior to its collection. In other communities, all recyclable materials are placed in a single bin for collection, and the sorting is handled later at a central facility. The latter method is known as "single-stream recycling."The most common consumer products recycled include aluminium such as beverages cans, copper such as wire, steel from food and aerosol cans, old steel furnishings or equipment, rubber tyres, polyethylene and PET bottles, glass bottles and jars, paperboard cartons, newspapers, magazines and light paper, and corrugated fiberboard boxes.

Waste not the Waste. Sign in Tamil Nadu, India

PVC, LDPE, PP, and PS are also recyclable. These items are usually composed of a single type of material, making them relatively easy to recycle into new products. The recycling of complex products (such as computers and electronic equipment) is more difficult, due to the additional dismantling and separation required.

Steel crushed and baled for recycling

The type of material accepted for recycling varies by city and country. Each city and country has different recycling programs in place that can handle the various types of recyclable materials. However, certain variation in acceptance is reflected in the resale value of the material once it is reprocessed.

Re-Use

Biological Reprocessing

An active compost heap.

Recoverable materials that are organic in nature, such as plant material, food scraps, and paper products, can be recovered through composting and digestion processes to decompose the organic matter. The resulting organic material is then recycled as mulch or compost for agricultural or landscaping purposes. In addition, waste gas from the process (such as methane) can be captured and used for generating electricity and heat (CHP/cogeneration) maximising efficiencies. The intention of biological processing in waste management is to control and ac-celerate the natural process of decomposition of organic matter.

Energy Recovery

Energy recovery from waste is the conversion of non-recyclable waste materials into usable heat, electricity, or fuel through a variety of processes, including combustion, gasification, pyrolyzation, anaerobic digestion, and landfill gas recovery. This process is often called waste-to-energy. Energy recovery from waste is part of the non-hazardous waste management hierarchy. Using energy recovery to convert non-recyclable waste materials into electricity and heat, generates a renewable energy source and can reduce carbon emissions by offsetting the need for energy from fossil sources as well as reduce methane generation from landfills. Globally, waste-to-energy accounts for 16% of waste management.

The energy content of waste products can be harnessed directly by using them as a direct combustion fuel, or indirectly by processing them into another type of fuel. Thermal treatment ranges from using waste as a fuel source for cooking or heating and the use of the gas fuel, to fuel for boilers to generate steam and electricity in a turbine. Pyrolysis and gasification are two related forms of thermal treatment where waste materials are heated to high temperatures with limited oxygen availability. The process usually occurs in a sealed vessel under high pressure. Pyrolysis of solid waste converts the material into solid, liquid and gas products. The liquid and gas can be burnt to produce energy or refined into other chemical products (chemical refinery). The solid residue (char) can be further refined into products such as activated carbon. Gasification and advanced Plasma arc gasification are used to convert organic materials directly into a synthetic gas (syngas) composed of carbon monoxide and hydrogen. The gas is then burnt to produce electricity and steam. An alternative to pyrolysis is high temperature and pressure supercritical water decomposition (hydrothermal monophasic oxidation).

Pyrolysis

Pyrolysis is a process of thermo-chemical decomposition of organic materials by heat in the absence of oxygen which produces various hydrocarbon gases. During pyrolysis, the molecules of object are subjected to very high temperatures leading to very high vibrations. Therefore, every molecule in the object is stretched and shaken to an extent that molecules starts breaking down. The rate of pyrolysis increases with temperature. In industrial applications, temperatures are above 430 °C (800 °F). Fast pyrolysis produces liquid fuel for feedstocks like wood. Slow pyrolysis produces gases and solid charcoal. Pyrolysis hold promise for conversion of waste biomass into useful liquid fuel. Pyrolysis of waste plastics can produce millions of litres of fuel. Solid products of this process contain metals, glass, sand and pyrolysis coke which cannot be converted to gas in the process.

Resource Recovery

Resource recovery is the systematic diversion of waste, which was intended for disposal, for a specific next use. It is the processing of recyclables to extract or recover materials and resources, or convert to energy. These activities are performed at a resource recovery facility. Resource recovery is not only environmentally important, but it is also cost effective. It decreases the amount of waste for disposal, saves space in landfills, and conserves natural resources.

Resource recovery (as opposed to waste management) uses LCA (life cycle analysis) attempts to offer alternatives to waste management. For mixed MSW (Municipal Solid Waste) a number of broad studies have indicated that administration, source separation and collection followed by reuse and recycling of the non-organic fraction and energy and compost/fertilizer production of the organic material via anaerobic digestion to be the favoured path.

As an example of how resource recycling can be beneficial, many of the items thrown away contain precious metals which can be recycled to create a profit, such as the components in circuit boards. Other industries can also benefit from resource recycling with the wood chippings in pallets and other packaging materials being passed onto sectors such as the horticultural profession. In this instance, workers can use the recycled chips to create paths, walkways, or arena surfaces.

Sustainability

The management of waste is a key component in a business' ability to maintaining ISO14001 accreditation. Companies are encouraged to improve their environmental efficiencies each year by eliminating waste through resource recovery practices, which are sustainability-related activities. One way to do this is by shifting away from waste management to resource recovery practices like recycling materials such as glass, food scraps, paper and cardboard, plastic bottles and metal. a lot of conferences will discuss this topic as the international Conference on Green Urabnism which will be held in Italy From 12–14 October 2016

Avoidance and Reduction Methods

An important method of waste management is the prevention of waste material being created, also known as waste reduction. Methods of avoidance include reuse of second-hand products, repairing broken items instead of buying new, designing products to be refillable or reusable (such as cotton instead of plastic shopping bags), encouraging consumers to avoid using disposable products (such as disposable cutlery), removing any food/liquid remains from cans and packaging, and designing products that use less material to achieve the same purpose (for example, lightweighting of beverage cans).

International Waste Movement

While waste transport within a given country falls under national regulations, trans-boundary movement of waste is often subject to international treaties. A major concern to many countries in the world has been hazardous waste. The Basel Convention, ratified by 172 countries, deprecates movement of hazardous waste from developed to less developed countries. The provisions of the Basel convention have been integrated into the EU waste shipment regulation. Nuclear waste, although considered hazardous, does not fall under the jurisdiction of the Basel Convention.

Benefits

Waste is not something that should be discarded or disposed of with no regard for future use. It can be a valuable resource if addressed correctly, through policy and practice. With rational and consistent waste management practices there is an opportunity to reap a range of benefits. Those benefits include:

1   Economic - Improving economic efficiency through the means of resource use, treatment and disposal and creating markets for recycles can lead to efficient practices in the production and consumption of products and materials resulting in valuable materials being recovered for reuse and the potential for new jobs and new business opportunities.

2   Social - By reducing adverse impacts on health by proper waste management practices, the resulting consequences are more appealing settlements. Better social advantages can lead to new sources of employment and potentially lifting communities out of poverty especially in some of the developing poorer countries and cities.

3   Environmental - Reducing or eliminating adverse impacts on the environmental through reducing, reusing and recycling, and minimizing resource extraction can provide improved air and water quality and help in the reduction of greenhouse gas emissions.

4   Inter-generational Equity - Following effective waste management practices can provide subsequent generations a more robust economy, a fairer and more inclusive society and a cleaner environment.

Challenges in Developing Countries

Waste management in cities with developing economies and economies in transition experience exhausted waste collection services, inadequately managed and uncontrolled dumpsites and the problems are worsening. Problems with governance also complicate the situation. Waste management, in these countries and cities, is an ongoing challenge and many struggle due to weak institutions, chronic under-resourcing and rapid urbanization. All of these challenges along with the lack of understanding of different factors that contribute to the hierarchy of waste management, affect the treatment of waste.

Technologies

Traditionally the waste management industry has been a late adopter of new technologies such as RFID (Radio Frequency Identification) tags, GPS and integrated software packages which enable better quality data to be collected without the use of estimation or manual data entry.

Bioconversion of Biomass to Mixed Alcohol Fuels

The bioconversion of biomass to mixed alcohol fuels can be accomplished using the MixAlco process. Through bioconversion of biomass to a mixed alcohol fuel, more energy from the biomass will end up as liquid fuels than in converting biomass to ethanol by yeast fermentation.

The process involves a biological/chemical method for converting any biodegradable material (e.g., urban wastes, such as municipal solid waste, biodegradable waste, and sewage sludge, agricultural residues such as corn stover, sugarcane bagasse, cotton gin trash, manure) into useful chemicals, such as carboxylic acids (e.g., acetic, propionic, butyric acid), ketones (e.g., acetone, methyl ethyl ketone, diethyl ketone) and biofuels, such as a mixture of primary alcohols (e.g., ethanol, propanol, n-butanol) and/or a mixture of secondary alcohols (e.g., isopropanol, 2-butanol, 3-pentanol). Because of the many products that can be economically produced, this process is a true biorefinery.

Pilot Plant (College Station, Texas)

The process uses a mixed culture of naturally occurring microorganisms found in natural habitats such as the rumen of cattle, termite guts, and marine and terrestrial swamps to anaerobically digest biomass into a mixture of carboxylic acids produced during the acidogenic and acetogenic stages of anaerobic digestion, however with the inhibition of the methanogenic final stage. The more popular methods for production of ethanol and cellulosic ethanol use enzymes that must be isolated first to be added to the biomass and thus convert the starch or cellulose into simple sugars, followed then by yeast fermentation into ethanol. This process does not need the addition of such enzymes as these microorganisms make their own.

As the microoganisms anaerobically digest the biomass and convert it into a mixture of carboxylic acids, the pH must be controlled. This is done by the addition of a buffering agent (e.g., ammonium bicarbonate, calcium carbonate), thus yielding a mixture of carboxylate salts. Methanogenesis, being the natural final stage of anaerobic digestion, is inhibited by the presence of the ammonium ions or by the addition of an inhibitor (e.g., iodoform). The resulting fermentation broth contains the produced carboxylate salts that must be dewatered. This is achieved efficiently by vapor-compression evaporation. Further chemical refining of the dewatered fermentation broth may then take place depending on the final chemical or biofuel product desired.

The condensed distilled water from the vapor-compression evaporation system is recycled back to the fermentation. On the other hand, if raw sewage or other waste water with high BOD in need of treatment is used as the water for the fermentation, the condensed distilled water from the evaporation can be recycled back to the city or to the original source of the high-BOD waste water. Thus, this process can also serve as a water treatment facility, while producing valuable chemicals or biofuels.

Because the system uses a mixed culture of microorganisms, besides not needing any enzyme addition, the fermentation requires no sterility or aseptic conditions, making this front step in the process more economical than in more popular methods for the production of cellulosic ethanol. These savings in the front end of the process, where volumes are large, allows flexibility for further chemical transformations after dewatering, where volumes are small.

Carboxylic Acids

Carboxylic acids can be regenerated from the carboxylate salts using a process known as "acid springing". This process makes use of a high-molecular-weight tertiary amine (e.g., trioctylamine), which is switched with the cation (e.g., ammonium or calcium). The resulting amine carboxylate can then be thermally decomposed into the amine itself, which is recycled, and the corresponding carboxylic acid. In this way, theoretically, no chemicals are consumed or wastes produced during this step.

Ketones

There are two methods for making ketones. The first one consists on thermally converting calcium carboxylate salts into the corresponding ketones. This was a common method for making acetone from calcium acetate during World War I. The other method for making ketones consists on converting the vaporized carboxylic acids on a catalytic bed of zirconium oxide.

Alcohols

Primary Alcohols

The undigested residue from the fermentation may be used in gasification to make hydrogen (H_2). This H_2 can then be used to hydrogenolyze the esters over a catalyst (e.g., copper chromite), which are produced by esterifying either the ammonium carboxylate salts (e.g., ammonium acetate, propionate, butyrate) or the carboxylic acids (e.g., acetic, propionic, butyric acid) with a high-molecular-weight alcohol (e.g., hexanol, heptanol). From the hydrogenolysis, the final products are the high-molecular-weight alcohol, which is recycled back to the esterification, and the corresponding primary alcohols (e.g., ethanol, propanol, butanol).

Secondary Alcohols

The secondary alcohols (e.g., isopropanol, 2-butanol, 3-pentanol) are obtained by hydrogenating over a catalyst (e.g., Raney nickel) the corresponding ketones (e.g., acetone, methyl ethyl ketone, diethyl ketone).

Drop-in Biofuels

The primary or secondary alcohols obtained as described above may undergo conversion to drop-in biofuels, fuels which are compatible with current fossil fuel infrastructure such as biogasoline, green diesel and bio-jet fuel. Such is done by subjecting the alcohols to dehydration followed by oligomerization using zeolite catalysts in a manner similar to the methanex process, which used to produce gasoline from methanol in New Zealand.

Acetic Acid Versus Ethanol

Cellulosic-ethanol manufacturing plants are bound to be net exporters of electricity because a large portion of the lignocellulosic biomass, namely lignin, remains undigested and it must be burned, thus producing electricity for the plant and excess electricity for the grid. As the market grows and this technology becomes more widespread, coupling the liquid fuel and the electricity markets will become more and more difficult.

Acetic acid, unlike ethanol, is biologically produced from simple sugars without the production of carbon dioxide:

$$C_6H_{12}O_6 \rightarrow 2\,CH_3CH_2OH + 2\,CO_2$$

(Biological production of ethanol)

$$C_6H_{12}O_6 \rightarrow 3\,CH_3COOH$$

(Biological production of acetic acid)

Because of this, on a mass basis, the yields will be higher than in ethanol fermentation. If then, the undigested residue (mostly lignin) is used to produce hydrogen by gasification, it is ensured that more energy from the biomass will end up as liquid fuels rather than excess heat/electricity.

$$3\,CH_3COOH + 6\,H_2 \rightarrow 3\,CH_3CH_2OH + 3\,H_2O$$

(Hydrogenation of acetic acid)

$$C_6H_{12}O_6 \text{ (from cellulose)} + 6\,H_2 \text{ (from lignin)} \rightarrow 3\,CH_3CH_2OH + 3\,H_2O$$

(Overall reaction)

A more comprehensive description of the economics of each of the fuels is given on the pages alcohol fuel and ethanol fuel, more information about the economics of various systems can be found on the central page biofuel.

Stage of Development

The system has been in development since 1991, moving from the laboratory scale (10 g/day) to the pilot scale (200 lb/day) in 2001. A small demonstration-scale plant (5 ton/day) has been constructed and is under operation and a 220 ton/day demonstration plant is expected in 2012.

Biofuel

A biofuel is a fuel that is produced through contemporary biological processes, such as agriculture and anaerobic digestion, rather than a fuel produced by geological processes such as those involved in the formation of fossil fuels, such as coal and petroleum, from prehistoric biological matter. Biofuels can be derived directly from plants, or indirectly from agricultural, commercial, domestic, and/or industrial wastes. Renewable biofuels generally involve contemporary carbon fixation, such as those that occur in plants or microalgae through the process of photosynthesis. Other renewable biofuels are made through the use or conversion of biomass (referring to recently living organisms, most often referring to plants or plant-derived materials). This biomass can be converted to convenient energy-containing substances in three different ways: thermal conversion, chemical conversion, and biochemical conversion. This biomass conversion can result in fuel in solid, liquid, or gas form. This new biomass can also be used directly for biofuels.

A bus fueled by biodiesel

Bioethanol is an alcohol made by fermentation, mostly from carbohydrates produced in sugar or starch crops such as corn, sugarcane, or sweet sorghum. Cellulosic biomass, derived from non-food sources, such as trees and grasses, is also being developed as a feedstock for ethanol production. Ethanol can be used as a fuel for vehicles in its pure form, but it is usually used as a gasoline additive to increase octane and improve vehicle emissions. Bioethanol is widely used in the USA and in Brazil. Current plant design does not provide for converting the lignin portion of plant raw materials to fuel components by fermentation.

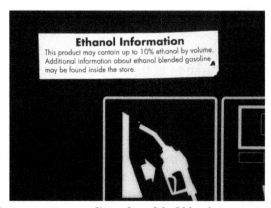

Information on pump regarding ethanol fuel blend up to 10%, California

Biodiesel can be used as a fuel for vehicles in its pure form, but it is usually used as a diesel additive to reduce levels of particulates, carbon monoxide, and hydrocarbons from diesel-powered vehicles. Biodiesel is produced from oils or fats using transesterification and is the most common biofuel in Europe.

In 2010, worldwide biofuel production reached 105 billion liters (28 billion gallons US), up 17% from 2009, and biofuels provided 2.7% of the world's fuels for road transport. Global ethanol fuel production reached 86 billion liters (23 billion gallons US) in 2010, with the United States and Brazil as the world's top producers, accounting together for 90% of global production. The world's largest biodiesel producer is the European Union, accounting for 53% of all biodiesel production in 2010. As of 2011, mandates for blending biofuels exist in 31 countries at the national level and in 29 states or provinces. The International Energy Agency has a goal for biofuels to meet more than a quarter of world demand for transportation fuels by 2050 to reduce dependence on petroleum and coal. The production of biofuels also led into a flourishing automotive industry, where by 2010, 79% of all cars produced in Brazil were made with a hybrid fuel system of bioethanol and gasoline.

There are various social, economic, environmental and technical issues relating to biofuels production and use, which have been debated in the popular media and scientific journals. These include: the effect of moderating oil prices, the "food vs fuel" debate, poverty reduction potential, carbon emissions levels, sustainable biofuel production, deforestation and soil erosion, loss of biodiversity, impact on water resources, rural social exclusion and injustice, shantytown migration, rural unskilled unemployment, and nitrous oxide (NO2) emissions.

Liquid Fuels for Transportation

Most transportation fuels are liquids, because vehicles usually require high energy density. This occurs naturally in liquids and solids. High energy density can also be provided by an internal combustion engine. These engines require clean-burning fuels. The fuels that are easiest to burn cleanly are typically liquids and gases. Thus, liquids meet the requirements of being both energy-dense and clean-burning. In addition, liquids (and gases) can be pumped, which means handling is easily mechanized, and thus less laborious.

First-Generation Biofuels

"First-generation" or conventional biofuels are made from sugar, starch, or vegetable oil.

Ethanol

Biologically produced alcohols, most commonly ethanol, and less commonly propanol and butanol, are produced by the action of microorganisms and enzymes through the fermentation of sugars or starches (easiest), or cellulose (which is more difficult). Biobutanol (also called biogasoline) is often claimed to provide a direct replacement for gasoline, because it can be used directly in a gasoline engine.

Neat ethanol on the left (A), gasoline on the right (G) at a filling station in Brazil

Ethanol fuel is the most common biofuel worldwide, particularly in Brazil. Alcohol fuels are produced by fermentation of sugars derived from wheat, corn, sugar beets, sugar cane, molasses and any sugar or starch from which alcoholic beverages such as whiskey, can be made (such as potato and fruit waste, etc.). The ethanol production methods used are enzyme digestion (to release sugars from stored starches), fermentation of the sugars, distillation and drying. The distillation process requires significant energy input for heat (sometimes unsustainable natural gas fossil fuel, but cellulosic biomass such as bagasse, the waste left after sugar cane is pressed to extract its juice, is the most common fuel in Brazil, while pellets, wood chips and also waste heat are more common in Europe) Waste steam fuels ethanol factory- where waste heat from the factories also is used in the district heating grid.

U.S. President George W. Bush looks at sugar cane, a source of biofuel, with Brazilian President Luiz Inácio Lula da Silva during a tour on biofuel technology at Petrobras in São Paulo, Brazil, 9 March 2007.

Ethanol can be used in petrol engines as a replacement for gasoline; it can be mixed with gasoline to any percentage. Most existing car petrol engines can run on blends of up to 15% bioethanol with petroleum/gasoline. Ethanol has a smaller energy density than that of gasoline; this means it takes more fuel (volume and mass) to produce the same amount of work. An advantage of ethanol (CH_3CH_2OH) is that it has a higher octane rating than ethanol-free gasoline available at roadside gas stations, which allows an increase of an engine's compression ratio for increased thermal efficiency. In high-altitude (thin air) locations, some states mandate a mix of gasoline and ethanol as a winter oxidizer to reduce atmospheric pollution emissions.

Ethanol is also used to fuel bioethanol fireplaces. As they do not require a chimney and are "flueless", bioethanol fires are extremely useful for newly built homes and apartments without a flue. The downsides to these fireplaces is that their heat output is slightly less than electric heat or gas fires, and precautions must be taken to avoid carbon monoxide poisoning.

Corn-to-ethanol and other food stocks has led to the development of cellulosic ethanol. According to a joint research agenda conducted through the US Department of Energy, the fossil energy ratios (FER) for cellulosic ethanol, corn ethanol, and gasoline are 10.3, 1.36, and 0.81, respectively.

Ethanol has roughly one-third lower energy content per unit of volume compared to gasoline. This is partly counteracted by the better efficiency when using ethanol (in a long-term test of more than 2.1 million km, the BEST project found FFV vehicles to be 1-26 % more energy efficient than petrol cars The BEST project), but the volumetric consumption increases by approximately 30%, so more fuel stops are required.

With current subsidies, ethanol fuel is slightly cheaper per distance traveled in the United States.

Biodiesel

Biodiesel is the most common biofuel in Europe. It is produced from oils or fats using transesterification and is a liquid similar in composition to fossil/mineral diesel. Chemically, it consists mostly of fatty acid methyl (or ethyl) esters (FAMEs). Feedstocks for biodiesel include animal fats, vegetable oils, soy, rapeseed, jatropha, mahua, mustard, flax, sunflower, palm oil, hemp, field pennycress, *Pongamia pinnata* and algae. Pure biodiesel (B100) currently reduces emissions with up to 60% compared to diesel Second generation B100.

Biodiesel can be used in any diesel engine when mixed with mineral diesel. In some countries, manufacturers cover their diesel engines under warranty for B100 use, although Volkswagen of Germany, for example, asks drivers to check by telephone with the VW environmental services department before switching to B100. B100 may become more viscous at lower temperatures, depending on the feedstock used. In most cases, biodiesel is compatible with diesel engines from 1994 onwards, which use 'Viton' (by DuPont) synthetic rubber in their mechanical fuel injection systems. Note however, that no vehicles are certified for using neat biodiesel before 2014, as there was no emission control protocol available for biodiesel before this date.

Electronically controlled 'common rail' and 'unit injector' type systems from the late 1990s onwards may only use biodiesel blended with conventional diesel fuel. These engines have finely metered and atomized multiple-stage injection systems that are very sensitive to the viscosity of the fuel. Many current-generation diesel engines are made so that they can run on B100 without altering the engine itself, although this depends on the fuel rail design. Since biodiesel is an effective solvent and cleans residues deposited by mineral diesel, engine filters may need to be replaced more often, as the biofuel dissolves old deposits in the fuel tank and pipes. It also effectively cleans the engine combustion chamber of carbon deposits, helping to maintain efficiency. In many European countries, a 5% biodiesel blend is widely used and is available at thousands of gas stations. Biodiesel is also an oxygenated fuel, meaning it contains a reduced amount of carbon and higher hydrogen and oxygen content than fossil diesel. This improves the combustion of biodiesel and reduces the particulate emissions from unburnt carbon. However, using neat biodiesel may increase NOx-emissions Nylund.N-O & Koponen.K. 2013. Fuel and Technology Alternatives for Buses. Overall Energy Efficiency and Emission Performance. IEA Bioenergy Task 46. Possibly the new emission standards Euro VI/EPA 10 will lead to reduced NOx-levels also when using B100.

Biodiesel is also safe to handle and transport because it is non-toxic and biodegradable, and has a high flash point of about 300 °F (148 °C) compared to petroleum diesel fuel, which has a flash point of 125 °F (52 °C).

In the USA, more than 80% of commercial trucks and city buses run on diesel. The emerging US biodiesel market is estimated to have grown 200% from 2004 to 2005. "By the end of 2006 biodiesel production was estimated to increase fourfold [from 2004] to more than" 1 billion US gallons (3,800,000 m³).

In France, biodiesel is incorporated at a rate of 8% in the fuel used by all French diesel vehicles. Avril Group produces under the brand Diester, a fifth of 11 million tons of biodiesel consumed annually by the European Union. It is the leading European producer of biodiesel.

Other Bioalcohols

Methanol is currently produced from natural gas, a non-renewable fossil fuel. In the future it is hoped to be produced from biomass as biomethanol. This is technically feasible, but the production is currently being postponed for concerns of Jacob S. Gibbs and Brinsley Coleberd that the economic viability is still pending. The methanol economy is an alternative to the hydrogen economy, compared to today's hydrogen production from natural gas.

Butanol (C_4H_9OH) is formed by ABE fermentation (acetone, butanol, ethanol) and experimental modifications of the process show potentially high net energy gains with butanol as the only liquid product. Butanol will produce more energy and allegedly can be burned "straight" in existing gasoline engines (without modification to the engine or car), and is less corrosive and less water-soluble than ethanol, and could be distributed via existing infrastructures. DuPont and BP are working together to help develop butanol. *E. coli* strains have also been successfully engineered to produce butanol by modifying their amino acid metabolism.

Green Diesel

Green diesel is produced through hydrocracking biological oil feedstocks, such as vegetable oils and animal fats. Hydrocracking is a refinery method that uses elevated temperatures and pressure in the presence of a catalyst to break down larger molecules, such as those found in vegetable oils, into shorter hydrocarbon chains used in diesel engines. It may also be called renewable diesel, hydrotreated vegetable oil or hydrogen-derived renewable diesel. Green diesel has the same chemical properties as petroleum-based diesel. It does not require new engines, pipelines or infrastructure to distribute and use, but has not been produced at a cost that is competitive with petroleum. Gasoline versions are also being developed. Green diesel is being developed in Louisiana and Singapore by ConocoPhillips, Neste Oil, Valero, Dynamic Fuels, and Honeywell UOP as well as Preem in Gothenburg, Sweden, creating what is known as Evolution Diesel.

Biofuel Gasoline

In 2013 UK researchers developed a genetically modified strain of Escherichia coli (E.Coli), which could transform glucose into biofuel gasoline that does not need to be blended. Later in 2013 UCLA researchers engineered a new metabolic pathway to bypass glycolysis and increase the rate of conversion of sugars into biofuel, while KAIST researchers developed a strain capable of producing short-chain alkanes, free fatty acids, fatty esters and fatty alcohols through the fatty acyl (acyl carrier protein (ACP)) to fatty acid to fatty acyl-CoA pathway *in vivo*. It is believed that in the future it will be possible to "tweak" the genes to make gasoline from straw or animal manure.

Vegetable Oil

Straight unmodified edible vegetable oil is generally not used as fuel, but lower-quality oil can and has been used for this purpose. Used vegetable oil is increasingly being processed into biodiesel, or (more rarely) cleaned of water and particulates and used as a fuel.

Filtered waste vegetable oil

As with 100% biodiesel (B100), to ensure the fuel injectors atomize the vegetable oil in the correct pattern for efficient combustion, vegetable oil fuel must be heated to reduce its viscosity to that of diesel, either by electric coils or heat exchangers. This is easier in warm or temperate climates. MAN B&W Diesel, Wärtsilä, and Deutz AG, as well as a number of smaller companies, such as Elsbett, offer engines that are compatible with straight vegetable oil, without the need for after-market modifications.

Walmart's truck fleet logs millions of miles each year, and the company planned to double the fleet's efficiency between 2005 and 2015. This truck is one of 15 based at Walmart's Buckeye, Arizona distribution center that was converted to run on a biofuel made from reclaimed cooking grease produced during food preparation at Walmart stores.

Vegetable oil can also be used in many older diesel engines that do not use common rail or unit injection electronic diesel injection systems. Due to the design of the combustion chambers in indirect injection engines, these are the best engines for use with vegetable oil. This system allows the relatively larger oil molecules more time to burn. Some older engines, especially Mercedes, are driven experimentally by enthusiasts without any conversion, a handful of drivers have experienced limited success with earlier pre-"Pumpe Duse" VW TDI engines and other similar engines with direct injection. Several companies, such as Elsbett or Wolf, have developed professional conversion kits and successfully installed hundreds of them over the last decades.

Oils and fats can be hydrogenated to give a diesel substitute. The resulting product is a straight-chain hydrocarbon with a high cetane number, low in aromatics and sulfur and does not contain oxygen. Hydrogenated oils can be blended with diesel in all proportions. They have several advantages over biodiesel, including good performance at low temperatures, no storage stability problems and no susceptibility to microbial attack.

Bioethers

Bioethers (also referred to as fuel ethers or oxygenated fuels) are cost-effective compounds that act as octane rating enhancers."Bioethers are produced by the reaction of reactive iso-olefins, such as iso-butylene, with bioethanol." Bioethers are created by wheat or sugar beet. They also enhance engine performance, whilst significantly reducing engine wear and toxic exhaust emissions. Though bioethers are likely to replace petroethers in the UK, it is highly unlikely they will become a fuel in and of itself due to the low energy density. Greatly reducing the amount of ground-level ozone emissions, they contribute to air quality.

When it comes to transportation fuel there are six ether additives- 1. Dimethyl Ether (DME) 2. Diethyl Ether (DEE) 3. Methyl Teritiary-Butyl Ether (MTBE) 4. Ethyl *ter*-butyl ether (ETBE) 5. *Ter*-amyl methyl ether (TAME) 6. *Ter*-amyl ethyl Ether (TAEE)

The European Fuel Oxygenates Association (aka EFOA) credits Methyl Tertiary-Butyl Ether (MTBE) and Ethyl ter-butyl ether (ETBE) as the most commonly used ethers in fuel to replace lead. Ethers were brought into fuels in Europe in the 1970s to replace the highly toxic compound. Although Europeans still use Bio-ether additives, the US no longer has an oxygenate requirement therefore bio-ethers are no longer used as the main fuel additive.

Biogas

Biogas is methane produced by the process of anaerobic digestion of organic material by anaerobes. It can be produced either from biodegradable waste materials or by the use of energy crops fed into anaerobic digesters to supplement gas yields. The solid byproduct, digestate, can be used as a biofuel or a fertilizer.

- Biogas can be recovered from mechanical biological treatment waste processing systems.

- Farmers can produce biogas from manure from their cattle by using anaerobic digesters.

Syngas

Syngas, a mixture of carbon monoxide, hydrogen and other hydrocarbons, is produced by partial combustion of biomass, that is, combustion with an amount of oxygen that is not sufficient to convert the biomass completely to carbon dioxide and water. Before partial combustion, the biomass is dried, and sometimes pyrolysed. The resulting gas mixture, syngas, is more efficient than direct combustion of the original biofuel; more of the energy contained in the fuel is extracted.

- Syngas may be burned directly in internal combustion engines, turbines or high-temperature fuel cells. The wood gas generator, a wood-fueled gasification reactor, can be connected to an internal combustion engine.

- Syngas can be used to produce methanol, DME and hydrogen, or converted via the Fischer-Tropsch process to produce a diesel substitute, or a mixture of alcohols that can be blended into gasoline. Gasification normally relies on temperatures greater than 700 °C.

- Lower-temperature gasification is desirable when co-producing biochar, but results in syngas polluted with tar.

Solid Biofuels

Examples include wood, sawdust, grass trimmings, domestic refuse, charcoal, agricultural waste, nonfood energy crops, and dried manure.

When raw biomass is already in a suitable form (such as firewood), it can burn directly in a stove or furnace to provide heat or raise steam. When raw biomass is in an inconvenient form (such as sawdust, wood chips, grass, urban waste wood, agricultural residues), the typical process is to densify the biomass. This process includes grinding the raw biomass to an appropriate particulate size (known as hogfuel), which, depending on the densification type, can be from 1 to 3 cm (0.4 to 1.2 in), which is then concentrated into a fuel product. The current processes produce wood pellets, cubes, or pucks. The pellet process is most common in Europe, and is typically a pure wood product. The other types of densification are larger in size compared to a pellet, and are compatible with a broad range of input feedstocks. The resulting densified fuel is easier to transport and feed into thermal generation systems, such as boilers.

Industry has used sawdust, bark and chips for fuel for decades, primary in the pulp and paper industry, and also bagasse (spent sugar cane) fueled boilers in the sugar cane industry. Boilers in the range of 500,000 lb/hr of steam, and larger, are in routine operation, using grate, spreader stoker, suspension burning and fluid bed combustion. Utilities generate power, typically in the range of 5 to 50 MW, using locally available fuel. Other industries have also installed wood waste fueled boilers and dryers in areas with low cost fuel.

One of the advantages of biomass fuel is that it is often a byproduct, residue or waste-product of other processes, such as farming, animal husbandry and forestry. In theory, this means fuel and food production do not compete for resources, although this is not always the case.

A problem with the combustion of raw biomass is that it emits considerable amounts of pollutants, such as particulates and polycyclic aromatic hydrocarbons. Even modern pellet boilers generate much more pollutants than oil or natural gas boilers. Pellets made from agricultural residues are usually worse than wood pellets, producing much larger emissions of dioxins and chlorophenols.

In spite of the above noted study, numerous studies have shown biomass fuels have significantly less impact on the environment than fossil based fuels. Of note is the US Department of Energy Laboratory, operated by Midwest Research Institute Biomass Power and Conventional Fossil Systems with and without CO2 Sequestration – Comparing the Energy Balance, Greenhouse Gas Emissions and Economics Study. Power generation emits significant amounts of greenhouse gases (GHGs), mainly carbon dioxide (CO_2). Sequestering CO_2 from the power plant flue gas can significantly reduce the GHGs from the power plant itself, but this is not the total picture. CO_2 capture and sequestration consumes additional energy, thus lowering the plant's fuel-to-electricity efficiency. To compensate for this, more fossil fuel must be procured and consumed to make up for lost capacity.

Taking this into consideration, the global warming potential (GWP), which is a combination of CO_2, methane (CH_4), and nitrous oxide (N_2O) emissions, and energy balance of the system need to be examined using a life cycle assessment. This takes into account the upstream processes which remain constant after CO_2 sequestration, as well as the steps required for additional power generation. Firing biomass instead of coal led to a 148% reduction in GWP.

A derivative of solid biofuel is biochar, which is produced by biomass pyrolysis. Biochar made from agricultural waste can substitute for wood charcoal. As wood stock becomes scarce, this alternative is gaining ground. In eastern Democratic Republic of Congo, for example, biomass briquettes are being marketed as an alternative to charcoal to protect Virunga National Park from deforestation associated with charcoal production.

Second-Generation (Advanced) Biofuels

Second generation biofuels, also known as advanced biofuels, are fuels that can be manufactured from various types of biomass. Biomass is a wide-ranging term meaning any source of organic carbon that is renewed rapidly as part of the carbon cycle. Biomass is derived from plant materials but can also include animal materials.

First generation biofuels are made from the sugars and vegetable oils found in arable crops, which can be easily extracted using conventional technology. In comparison, second generation biofuels are made from lignocellulosic biomass or woody crops, agricultural residues or waste, which makes it harder to extract the required fuel. A series of physical and chemical treatments might be required to convert lignocellulosic biomass to liquid fuels suitable for transportation.

Sustainable Biofuels

Biofuels in the form of liquid fuels derived from plant materials, are entering the market, driven mainly by the perception that they reduce climate gas emissions, and also by factors such as oil price spikes and the need for increased energy security. However, many of the biofuels that are currently being supplied have been criticised for their adverse impacts on the natural environment, food security, and land use. In 2008, the Nobel-prize winning chemist Paul J. Crutzen published findings that the release of nitrous oxide (N_2O) emissions in the production of biofuels means that overall they contribute more to global warming than the fossil fuels they replace.

The challenge is to support biofuel development, including the development of new cellulosic technologies, with responsible policies and economic instruments to help ensure that biofuel commercialization is sustainable. Responsible commercialization of biofuels represents an opportunity to enhance sustainable economic prospects in Africa, Latin America and Asia.

According to the Rocky Mountain Institute, sound biofuel production practices would not hamper food and fibre production, nor cause water or environmental problems, and would enhance soil fertility. The selection of land on which to grow the feedstocks is a critical component of the ability of biofuels to deliver sustainable solutions. A key consideration is the minimisation of biofuel competition for prime cropland.

Biofuels by Region

There are international organizations such as IEA Bioenergy, established in 1978 by the OECD International Energy Agency (IEA), with the aim of improving cooperation and information exchange between countries that have national programs in bioenergy research, development and deployment. The UN International Biofuels Forum is formed by Brazil, China, India, Pakistan, South Africa, the United States and the European Commission. The world leaders in biofuel development and use are Brazil, the United States, France, Sweden and Germany. Russia also has 22% of world's forest, and is a big biomass (solid biofuels) supplier. In 2010, Russian pulp and paper maker, Vyborgskaya Cellulose, said they would be producing pellets that can be used in heat and electricity generation from its plant in Vyborg by the end of the year. The plant will eventually produce about 900,000 tons of pellets per year, making it the largest in the world once operational.

Bio Diesel Powered Fast Attack Craft Of Indian Navy patrolling during IFR 2016.The green bands on the vessels are indicative of the fact that the vessels are powered by bio-diesel

Biofuels currently make up 3.1% of the total road transport fuel in the UK or 1,440 million litres. By 2020, 10% of the energy used in UK road and rail transport must come from renewable sources – this is the equivalent of replacing 4.3 million tonnes of fossil oil each year. Conventional biofuels are likely to produce between 3.7 and 6.6% of the energy needed in road and rail transport, while advanced biofuels could meet up to 4.3% of the UK's renewable transport fuel target by 2020.

Air Pollution

Biofuels are different from fossil fuels in regard to greenhouse gases but are similar to fossil fuels in that biofuels contribute to air pollution. Burning produces airborne carbon particulates, carbon monoxide and nitrous oxides. The WHO estimates 3.7 million premature deaths worldwide in 2012 due to air pollution. Brazil burns significant amounts of ethanol biofuel. Gas chromatograph studies were performed of ambient air in São Paulo, Brazil, and compared to Osaka, Japan, which does not burn ethanol fuel. Atmospheric Formaldehyde was 160% higher in Brazil, and Acetaldehyde was 260% higher.

Debates Regarding the Production and Use of Biofuel

There are various social, economic, environmental and technical issues with biofuel production and use, which have been discussed in the popular media and scientific journals. These include: the effect of moderating oil prices, the "food vs fuel" debate, food prices, poverty reduction potential, energy ratio, energy requirements, carbon emissions levels, sustainable biofuel production, deforestation and soil erosion, loss of biodiversity, impact on water resources, the possible modifications necessary to run the engine on biofuel, as well as energy balance and efficiency. The International Resource Panel, which provides independent scientific assessments and expert advice on a variety of resource-related themes, assessed the issues relating to biofuel use in its first report *Towards sustainable production and use of resources: Assessing Biofuels*. "Assessing Biofuels" outlined the wider and interrelated factors that need to be considered when deciding on the relative merits of pursuing one biofuel over another. It concluded that not all biofuels perform equally in terms of their impact on climate, energy security and ecosystems, and suggested that environmental and social impacts need to be assessed throughout the entire life-cycle.

Another issue with biofuel use and production is the US has changed mandates many times because the production has been taking longer than expected. The Renewable Fuel Standard (RFS) set by congress for 2010 was pushed back to at best 2012 to produce 100 million gallons of pure ethanol (not blended with a fossil fuel).

Current Research

Research is ongoing into finding more suitable biofuel crops and improving the oil yields of these crops. Using the current yields, vast amounts of land and fresh water would be needed to produce enough oil to completely replace fossil fuel usage. It would require twice the land area of the US to be devoted to soybean production, or two-thirds to be devoted to rapeseed production, to meet current US heating and transportation needs.

Specially bred mustard varieties can produce reasonably high oil yields and are very useful in crop rotation with cereals, and have the added benefit that the meal left over after the oil has been pressed out can act as an effective and biodegradable pesticide.

The NFESC, with Santa Barbara-based Biodiesel Industries, is working to develop biofuels technologies for the US navy and military, one of the largest diesel fuel users in the world. A group of Spanish developers working for a company called Ecofasa announced a new biofuel made from trash. The fuel is created from general urban waste which is treated by bacteria to produce fatty acids, which can be used to make biofuels.

Ethanol Biofuels

As the primary source of biofuels in North America, many organizations are conducting research in the area of ethanol production. The National Corn-to-Ethanol Research Center (NCERC) is a research division of Southern Illinois University Edwardsville dedicated solely to ethanol-based biofuel research projects. On the federal level, the USDA conducts a large amount of research regarding ethanol production in the United States. Much of this research is targeted toward the effect of ethanol production on domestic food markets. A division of the U.S. Department of Energy, the National Renewable Energy Laboratory (NREL), has also conducted various ethanol research projects, mainly in the area of cellulosic ethanol.

Cellulosic ethanol commercialization is the process of building an industry out of methods of turning cellulose-containing organic matter into fuel. Companies, such as Iogen, POET, and Abengoa, are building refineries that can process biomass and turn it into bioethanol. Companies, such as Diversa, Novozymes, and Dyadic, are producing enzymes that could enable a cellulosic ethanol future. The shift from food crop feedstocks to waste residues and native grasses offers significant opportunities for a range of players, from farmers to biotechnology firms, and from project developers to investors.

As of 2013, the first commercial-scale plants to produce cellulosic biofuels have begun operating. Multiple pathways for the conversion of different biofuel feedstocks are being used. In the next few years, the cost data of these technologies operating at commercial scale, and their relative performance, will become available. Lessons learnt will lower the costs of the industrial processes involved.

In parts of Asia and Africa where drylands prevail, sweet sorghum is being investigated as a potential source of food, feed and fuel combined. The crop is particularly suitable for growing in arid conditions, as it only extracts one seventh of the water used by sugarcane. In India, and other places, sweet sorghum stalks are used to produce biofuel by squeezing the juice and then fermenting into ethanol.

A study by researchers at the International Crops Research Institute for the Semi-Arid Tropics (ICRISAT) found that growing sweet sorghum instead of grain sorghum could increase farmers incomes by US$40 per hectare per crop because it can provide fuel in addition to food and animal feed. With grain sorghum currently grown on over 11 million hectares (ha) in Asia and on 23.4 million ha in Africa, a switch to sweet sorghum could have a considerable economic impact.

Algae Biofuels

From 1978 to 1996, the US NREL experimented with using algae as a biofuels source in the "Aquatic Species Program". A self-published article by Michael Briggs, at the UNH Biofuels Group, offers estimates for the realistic replacement of all vehicular fuel with biofuels by using algae that have a natural oil content greater than 50%, which Briggs suggests can be grown on algae ponds at wastewater treatment plants. This oil-rich algae can then be extracted from the system and processed into biofuels, with the dried remainder further reprocessed to create ethanol. The production of algae to harvest oil for biofuels has not yet been undertaken on a commercial scale, but feasibility studies have been conducted to arrive at the above yield estimate. In addition to its projected high yield, algaculture — unlike crop-based biofuels — does not entail a decrease in food production, since it requires neither farmland nor fresh water. Many companies are pursuing algae bioreactors for various purposes, including scaling up biofuels production to commercial levels. Prof. Rodrigo E. Teixeira from the University of Alabama in Huntsville demonstrated the extraction of biofuels lipids from wet algae using a simple and economical reaction in ionic liquids.

Jatropha

Several groups in various sectors are conducting research on *Jatropha curcas*, a poisonous shrub-like tree that produces seeds considered by many to be a viable source of biofuels feedstock oil. Much of this research focuses on improving the overall per acre oil yield of Jatropha through advancements in genetics, soil science, and horticultural practices.

SG Biofuels, a San Diego-based jatropha developer, has used molecular breeding and biotechnology to produce elite hybrid seeds that show significant yield improvements over first-generation varieties. SG Biofuels also claims additional benefits have arisen from such strains, including improved flowering synchronicity, higher resistance to pests and diseases, and increased cold-weather tolerance.

Plant Research International, a department of the Wageningen University and Research Centre in the Netherlands, maintains an ongoing Jatropha Evaluation Project that examines the feasibility of large-scale jatropha cultivation through field and laboratory experiments. The Center for Sustainable Energy Farming (CfSEF) is a Los Angeles-based nonprofit research organization dedicated to jatropha research in the areas of plant science, agronomy, and horticulture. Successful exploration of these disciplines is projected to increase jatropha farm production yields by 200-300% in the next 10 years.

Fungi

A group at the Russian Academy of Sciences in Moscow, in a 2008 paper, stated they had isolated large amounts of lipids from single-celled fungi and turned it into biofuels in an economically efficient manner. More research on this fungal species, *Cunninghamella japonica*, and others, is likely to appear in the near future. The recent discovery of a variant of the fungus *Gliocladium roseum* (later renamed Ascocoryne sarcoides) points toward the production of so-called myco-diesel from cellulose. This organism was recently discovered in the rainforests of northern Patagonia, and has the unique capability of converting cellulose into medium-length hydrocarbons typically found in diesel fuel. Many other fungi that can degrade cellulose and other polymers have been observed to produce molecules that are currently being engineered using organisms from other kingdoms, suggesting that fungi may play a large role in the bio-production of fuels in the future (reviewed in).

Animal Gut Bacteria

Microbial gastrointestinal flora in a variety of animals have shown potential for the production of biofuels. Recent research has shown that TU-103, a strain of *Clostridium* bacteria found in Zebra feces, can convert nearly any form of cellulose into butanol fuel. Microbes in panda waste are being investigated for their use in creating biofuels from bamboo and other plant materials. There has also been substantial research into the technology of using the gut microbiomes of wood-feeding insects for the conversion of lignocellulotic material into biofuel.

Greenhouse Gas Emissions

Some scientists have expressed concerns about land-use change in response to greater demand for crops to use for biofuel and the subsequent carbon emissions. The payback period, that is, the time it will take biofuels to pay back the carbon debt they acquire due to land-use change, has been estimated to be between 100 and 1000 years, depending on the specific instance and location of land-use change. However, no-till practices combined with cover-crop practices can reduce the payback period to three years for grassland conversion and 14 years for forest conversion.

A study conducted in the Tocantis State, in northern Brazil, found that many families were cutting down forests in order to produce two conglomerates of oilseed plants, the J. curcas (JC group) and the R. communis (RC group). This region is composed of 15% Amazonian rainforest with high biodiversity, and 80% Cerrado forest with lower biodiversity. During the study, the farmers that planted the JC group released over 2193 Mg CO_2, while losing 53-105 Mg CO_2 sequestration from deforestation; and the RC group farmers released 562 Mg CO_2, while losing 48-90 Mg CO_2 to be sequestered from forest depletion. The production of these types of biofuels not only led into an increased emission of carbon dioxide, but also to lower efficiency of forests to absorb the gases that these farms were emitting. This has to do with the amount of fossil fuel the production of fuel crops involves. In addition, the intensive use of monocropping agriculture requires large amounts of water irrigation, as well as of fertilizers, herbicides and pesticides. This does not only lead to poisonous chemicals to disperse on water runoff, but also to the emission of nitrous oxide (NO_2) as a fertilizer by-product, which is three hundred times more efficient in producing a greenhouse effect than carbon dioxide (CO_2).

Converting rainforests, peatlands, savannas, or grasslands to produce food crop–based biofuels in Brazil, Southeast Asia, and the United States creates a "biofuel carbon debt" by releasing 17 to 420 times more CO_2 than the annual greenhouse gas (GHG) reductions that these biofuels would provide by displacing fossil fuels. Biofuels made from waste biomass or from biomass grown on abandoned agricultural lands incur little to no carbon debt.

Water Use

In addition to water required to grow crops, biofuel facilities require significant process water.

Biohydrogen

Biohydrogen is defined as hydrogen produced biologically, most commonly by algae, bacteria and archaea. Biohydrogen is a potential biofuel obtainable from both cultivation and from waste organic materials.

Microbial hydrogen production.

Introduction

Currently, there is a huge demand for hydrogen. There is no log of the production volume and use of hydrogen world-wide, however consumption of hydrogen was estimated to have reached 900 billion cubic meters in 2011.

Refineries are large-volume producers and consumers of hydrogen. Today 96% of all hydrogen is derived from fossil fuels, with 48% from natural gas, 30% from hydrocarbons, 18% from coal and about 4% from electrolysis. Oil-sands processing, gas-to-liquids and coal gasification projects that are ongoing, require a huge amount of hydrogen and is expected to boost the requirement significantly within the next few years. Environmental regulations implemented in most countries, increase the hydrogen requirement at refineries for gas-line and diesel desulfurization.

An important future application of hydrogen could be as an alternative for fossil fuels, once the oil deposits are depleted. This application is however dependent on the development of storage techniques to enable proper storage, distribution and combustion of hydrogen. If the cost of hydrogen production, distribution, and end-user technologies decreases, hydrogen as a fuel could be entering the market in 2020.

Industrial fermentation of hydrogen, or whole-cell catalysis, requires a limited amount of energy, since fission of water is achieved with whole cell catalysis, to lower the activation energy. This allows hydrogen to be produced from any organic material that can be derived through whole cell catalysis since this process does not depend on the energy of substrate.

Algal Biohydrogen

In 1939 a German researcher named Hans Gaffron, while working at the University of Chicago, observed that the alga he was studying, *Chlamydomonas reinhardtii* (a green alga), would sometimes switch from the production of oxygen to the production of hydrogen. Gaffron never discovered the cause for this change and for many years other scientists failed in their attempts at its discovery. In the late 1990s professor Anastasios Melis a researcher at the University of California at Berkeley discovered that if the algal culture medium is deprived of sulfur it will switch from the production of oxygen (normal photosynthesis), to the production of hydrogen. He found that the enzyme responsible for this reaction is hydrogenase, but that the hydrogenase lost this function in the presence of oxygen. Melis found that depleting the amount of sulfur available to the algae interrupted its internal oxygen flow, allowing the hydrogenase an environment in which it can react, causing the algae to produce hydrogen. *Chlamydomonas moewusii* is also a good strain for the production of hydrogen. Scientists at the U.S. Department of Energy's Argonne National Laboratory are currently trying to find a way to take the part of the hydrogenase enzyme that creates the hydrogen gas and introduce it into the photosynthesis process. The result would be a large amount of hydrogen gas, possibly on par with the amount of oxygen created.

It would take about 25,000 square kilometres to be sufficient to displace gasoline use in the US. To put this in perspective, this area represents approximately 10% of the area devoted to growing soya in the US. The US Department of Energy has targeted a selling price of $2.60 / kg as a goal for making renewable hydrogen economically viable. 1 kg is approximately the energy equivalent to a gallon of gasoline. To achieve this, the efficiency of light-to-hydrogen conversion must reach 10% while current efficiency is only 1% and selling price is estimated at $13.53 / kg. According to the DOE cost estimate, for a refueling station to supply 100 cars per day, it would need 300 kg. With current technology, a 300 kg per day stand-alone system will require 110,000 m^2 of pond area, 0.2 g/l cell concentration, a truncated antennae mutant and 10 cm pond depth. Areas of research to increase efficiency include developing oxygen-tolerant FeFe-hydrogenases and increased hydrogen production rates through improved electron transfer.

Bacterial Biohydrogen

Process Requirements

If hydrogen by fermentation is to be introduced as an industry, the fermentation process will be dependent on organic acids as substrate for photo-fermentation. The organic acids are necessary for high hydrogen production rates.

The organic acids can be derived from any organic material source such as sewage waste waters or agricultural wastes. The most important organic acids are acetic acid (HAc), butyric acid (HBc) and propionic acid (HPc). A huge advantage is that production of hydrogen by fermentation does not require glucose as substrate.

The fermentation of hydrogen has to be a continuous fermentation process, in order sustain high production rates, since the amount of time for the fermentation to enter high production rates are in days.

Fermentation

Several strategies for the production of hydrogen by fermentation in lab-scale have been found in literature. However no strategies for industrial-scale productions have been found. In order to define an industrial-scale production, the information from lab-scale experiments has been scaled to an industrial-size production on a theoretical basis. In general, the method of hydrogen fermentation is referred to in three main categories. The first category is dark-fermentation, which is fermentation which does not involve light. The second category is photo-fermentation, which is fermentation which requires light as the source of energy. The third is combined-fermentation, which refers to the two fermentations combined.

Dark Fermentation

There are several bacteria with a potential for hydrogen production. The Gram-positive bacteria of the *Clostridium* genus, is promising because it has a natural high hydrogen production rate. In addition, it is fast growing and capable of forming endospores, which make the bacteria easy to handle in industrial application.

Species of the *Clostridium* genus allow hydrogen production in mixed cultures, under mesophilic or thermophilic conditions within a pH range of 5.0 to 6.5. Dark-fermentation with mixed cultures seems promising since a mixed bacterial environment within the fermenter, allows cooperation of different species to efficiently degrade and convert organic waste materials into hydrogen, accompanied by the formation of organic acids. The clostridia produce H2 via a reversible hydrogenase (H2ase) enzyme (2H + 2e <=> H2) and this reaction is important in achieving the redox balance of fermentation. The rate of H2 formation is inhibited as H2 production causes the partial pressure of H2 (pH2) to increase. This can limit substrate conversion and growth and the bacteria may respond by switching to a different metabolic pathway in order to achieve redox balance, energy generation and growth - by producing solvents instead of hydrogen and organic acids.

Enteric bacteria such as *Escherichia coli* and *Enterobacter aerogenes* are also interesting for biohydrogen fermentation.) Dissecting the roles of E. coli hydrogenases in biohydrogen production. Unlike the clostridia, the enteric bacteria produce hydrogen primarily (or exclusively in the case of *E. coli*) by the cleavage of formate ($HCOOH \longrightarrow H_2 + CO_2$), which serves to detoxify the medium by removing formate. Cleavage is not a redox reaction and it has no consequence on the redox balance of fermentation. This detoxification is particularly important for *E. coli* as it cannot protect itself by forming endospores. Formate cleavage is an irreversible reaction, hence H2 production is insensitive to the partial pressure of hydrogen (pH2) in the fermenter.

E. coli has been referred to as the workhorse of molecular microbiology and many workers have investigated metabolic engineering approaches to improve biohydrogen fermentation in *E. coli*.

Whereas oxygen kills clostridia, the enteric bacteria are facultative anaerobes; they grow very quickly when oxygen is available and transition progressively from aerobic to anaerobic metabolism as oxygen becomes depleted. Growth rate is much slower during anaerobic fermentation than during aerobic respiration because fermentation less metabolic energy from the same substrate. In practical terms, facultative anaerobes are useful because they can be grown quickly to a very high concentration with oxygen and then used to produce hydrogen at a high rate when the oxygen supply is stopped.

For fermentation to be sustainable at industrial-scale, it is necessary to control the bacterial community inside the fermenter. Feedstocks may contain micro-organisms, which could cause changes in the microbial community inside the fermenter. The enteric bacteria and most clostridia are mesophilic; they have an optimum temperature of around 30 degrees C as do many common environmental microorganisms. Therefore, these fermentations are susceptible to changes in the microbial community unless the feedstock is sterilised, for example where a hydrothermal pretreatment is involved, sterilisation is a side-effect. A way to prevent harmful micro-organisms from gaining control of the bacterial environment inside the fermenter could be through addition of the desired bacteria. Hyperthermophilic archaea such as *Thermotoga neapolitana* can also be used for hydrogen fermentation. Because they operate at around 70 degrees C, there is little chance of feedstock contaminants becoming established.

Fermentations produce organic acids are toxic to the bacteria. High concentrations inhibit the fermentation process and may trigger changes in metabolism and resistance mechanisms such as sporulation in different species. This fermentation of hydrogen is accompanied production of carbon-dioxide which can be separated from hydrogen with a passive separation process.

The fermentation will convert some of the substrate (e.g. waste) into biomass instead of hydrogen. The biomass is, however, a carbohydrate-rich by-product which can be fed back into the fermenter, to ensure that the process is sustainable. Fermentation of hydrogen by dark-fermentation is restricted by incomplete degradation of organic material, into organic acids and this is why we need the photo-fermentation.

The separation of organic acids from biomass in the outlet stream can be done with a settler tank in the outlet stream, where the sludge (biomass) is pumped back into the fermenter to increase the rate of hydrogen production.

In traditional fermentation systems, the dilution rate must be carefully controlled as it affects the concentration of bacterial cells and toxic end-products (organic acids and solvents) inside the fermenter. A more complex *electro-fermentation* technique decouples the retention of water and biomass and overcomes inhibition by organic acids.

Photo-Fermentation

Photo-fermentation refers to the method of fermentation where light is required as the source of energy. This fermentation relies on photosynthesis to maintain the cellular energy levels. Fermentation by photosynthesis compared to other fermentations has the advantage of light as the source of energy instead of sugar. Sugars are usually available in limited quantities.

All plants, algae and some bacteria are capable of photosynthesis: utilizing light as the source of metabolic energy. *Cyanobacteria* are frequently mentioned capable of hydrogen production by oxygenic photosynthesis. However the purple non-sulphur (PNS) bacteria (e.g. genus *Rhodobacter*) hold significant promise for the production of hydrogen by anoxygenic photosynthesis and photo-fermentation.

Studies have shown that *Rhodobacter sphaeroides* is highly capable of hydrogen production while feeding on organic acids, consuming 98% to 99% of the organic acids during hydrogen production. Organic acids may be sourced sustainably from the dark fermentation of waste feedstocks. The resultant system is called *combined fermentation*.

Photo-fermentative bacteria can use light in the wavelength range 400-1000 nm (visible and near-infrared) which differs from algae and cyanobacteria (400-700 nm; visible).

Currently there is limited experience with photo-fermentation at industrial-scale. The distribution of light within the industrial scale photo-fermenter has to be designed to minimise self-shading. Therefore, any externally illuminated photobioreactor must have a high ratio of high surface area to volume. As a result, photobioreactor construction is materials-intensive and expensive.

A method to ensure proper light distribution and limit self-shading within the fermenter, could be to distribute the light with an optic fiber where light is transferred into the fermenter and distributed from within the fermenter. Photo-fermentation with *Rhodobacter sphaeroides* require mesophilic conditions. An advantage of the optical fiber photobioreactor is that radiant heat-gain can be controlled by dumping excess light and filtering out wavelengths which cannot be used by the organisms.

Combined Fermentation

Combining dark- and photo-fermentation has shown to be the most efficient method to produce hydrogen through fermentation. The combined fermentation allows the organic acids produced during dark-fermentation of waste materials, to be used as substrate in the photo-fermentation process. Many independent studies show this technique to be effective and practical.

For industrial fermentation of hydrogen to be economical feasible, by-products of the fermentation process has to be minimized. Combined fermentation has the unique advantage of allowing reuse of the otherwise useless chemical, organic acids, through photosynthesis.

Many wastes are suitable for fermentation and this is equivalent the initial stages of anaerobic digestion, now the most important biotechnology for energy from waste. One of the main challenges in combined fermentation is that effluent fermentation contains not only useful oroganic acids but excess nitrogenous compounds and ammonia, which inhibit nitrogenase activity by wild-type PNS bacteria. The problem can be solved by genetic engineering to interrupt down-regulation of nitrogenase in response to nitrogen excess. However, genetically engineered bacterial strains may pose containment issues for application. A physical solution to this problem was developed at The University of Birmingham UK, which involves selective electro-separation of organic acids from an active fermentation. The energetic cost of electro-separation of organic acids was found to be acceptable in a combined fermentation. "Electro-fermentation" has the side-effect of a continuous, high-rate dark hydrogen fermentation.

As the method for hydrogen production, combined fermentation currently holds significant promise.

Metabolic Processes

The metabolic process for hydrogen production are dependent on the reduction of the metabolite ferredoxin (except in the enteric bacteria, where an alternative formate pathway operates).

$$4H^+ + 4 \text{ ferredoxin(red)} \rightarrow 4 \text{ ferredoxin(ox)} + 2 H_2$$

For this process to run, ferredoxin has to be recycled through oxidation. The recycling process is dependent on the transfer of electrons from nicotinamide adenine dinucleotide (NADH) to ferredoxin.

$$2 \text{ ferredoxin(ox)} + NADH_2 \rightarrow 2 \text{ ferredoxin(red)} + 2H^+ + NAD^+$$

The enzymes that catalyse this recycling process are referred to as hydrogen-forming enzymes and have complex metalloclusters in their active site and require several maturation proteins to attain their active form. The hydrogen-forming enzymes are inactivated by molecular oxygen and must be separated from oxygen, to produce hydrogen.

The three main classes of hydrogen-forming enzymes are [FeFe]-hydrogenase, [NiFe]-hydrogenase and nitrogenase. These enzymes behave differently in dark-fermentation with *Clostridium* and photo-fermentation with *Rhodobacter*. The interplay of these enzymes are the key in hydrogen production by fermentation.

Clostridium

The interplay of the hydrogen-forming enzymes in *Clostridium* is unique with little or no involvement of nitrogenase. The hydrogen production in this bacteria is mostly due to [FeFe]-hydrogenase, which activity is a hundred times higher than [NiFe]-hydrogenase and a thousand times higher than nitrogenase. [FeFe]-hydrogenase has a Fe-Fe catalytic core with a variety of electron donors and acceptors.

The enzyme [NiFe]-hydrogenase in *Clostridium*, catalyse a reversible oxidation of hydrogen. [NiFe]-hydrogenase is responsible for hydrogen uptake, utilizing the electrons from hydrogen for cellular maintenance.

In Clostridium, glucose is broken down into pyruvate and nicotinamide adenine dinucleotide (NADH). The formed pyruvate is then further converted to acetyl-CoA and hydrogen by pyruvate ferredoxin oxidoreductase with the reduction of ferredoxin. Acetyl-CoA is then converted to acetate, butyrate and propionate.

Acetate fermentation processes are well understood and have a maximum yield of 4 mol hydrogen pr. mol glucose. The yield of hydrogen from the conversion of acetyl-CoA to butyrate, has half the yield as the conversion to acetate. In mixed cultures of *Clostridium* the reaction is a combined production of acetate, butyrate and propionate. The organic acids which are the by-product of fermentation with *Clostridium*, can be further processed as substrate for hydrogen production with *Rhodobacter*.

Rhodobacter

The purple non-sulphur (PNS) bacteria *Rhodobacter sphaeroides* is able to produce hydrogen from a wide range of organic compounds (chiefly organic acids) and light.

The photo-system required for hydrogen production in *Rhodobacter* (PS-I), differ from its oxygenic photosystem (PS-II) due to the requirement of organic acids and the inability to oxidize water. In the absence of water-splitting photosynthesis is anoxygenic. Therefore, hydrogen production is sustained without inhibition from generated oxygen.

In PNS bacteria, hydrogen production is due to catalysis by nitrogenase. Hydrogenases are also present but the production of hydrogen by [FeFe]-hydrogenase is less than 10 times the hydrogen uptake by [NiFe]-hydrogenase.

Only under nitrogen-deficient conditions is nitrogenase activity sufficient to overcome uptake hydrogenase activity, resulting in net generation of hydrogen.

The main photosynthetic membrane complex is PS-I which accounts for most of the light-harvest. The photosynthetic membrane complex PS-II produces oxygen, which inhibit hydrogen production and thus low partial pressures of oxygen most be sustained during fermentation.

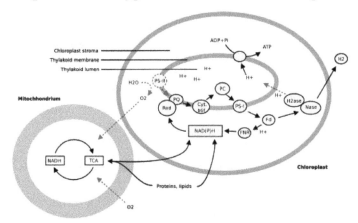

Rhodobacter hydrogen metabolism

The range of photosynthetically active radiation for PNS bacteria is 400-1000 nm. This includes the visible (VIS) and near-infrared (NIR)sections of the spectrum and not (despite erroneous writings) ultraviolet. This range is wider than that of algae and cyanobacteria (400-700 nm; VIS). The response to light (action spectrum) varies dramatically across the active range. Around 80% of activity is associated with the NIR. VIS is absorbed but much less efficiently utilised.

To attain high production rates of hydrogen, the hydrogen production by nitrogenase has to exceed the hydrogen uptake by hydrogenase. The substrate is oxidized through the tricarboxylic acids circle and the produced electrons are transferred to the nitrogenase catalysed reduction of protons to hydrogen, through the electron transport chain.

LED-Fermenter

To build an industrial-size photo-fermenter without using large areas of land could achieved using a fermenter with light-emitting diodes (LED) as light source. This design prevents self-shading within the fermenter, require limited energy to maintain photosynthesis and has very low installation costs. This design would also allow cheap models to be built for educational purpose.

However, it is impossible for any photobioreactor using artificial lights to generate energy. The maximum light conversion efficiency into hydrogen is about 10% (by PNS bacteria) and the maximum efficiency of electricity generation from hydrogen about 80% (by PEM fuel cell) and the maximum efficiency of light generation from electricity (via LED) is about 80%. This represents a cycle of diminishing returns. For the purposes of fuel or energy production sunlight is necessary but artificially lit photobioreactors such as the LED-fermenter could be useful for the production of other valuable commodities.

Metabolic Engineering

There is a huge potential for improving hydrogen yield by metabolic engineering. The bacteria *Clostridium* could be improved for hydrogen production by disabling the uptake hydrogenase, or disabling the oxygen system. This will make the hydrogen production robust and increase the hydrogen yield in the dark-fermentation step.

The photo-fermentation step with Rhodobacter, is the step which is likely to gain the most from metabolic engineering. An option could be to disable the uptake-hydrogenase or to disable the photosynthetic membrane system II (PS-II). Another improvement could be to decrease the expression of pigments, which shields of the photo-system.

Digestate

Digestate is the material remaining after the anaerobic digestion of a biodegradable feedstock. Anaerobic digestion produces two main products: digestate and biogas. Digestate is produced both by acidogenesis and methanogenesis and each has different characteristics.

Acidogenic digestate produced from mixed municipal waste

Acidogenic Digestate

Acidogenic digestate is fibrous and consists of structural plant matter including lignin and cellulose. Acidogenic digestate has high moisture retention properties. The digestate may also contain minerals and remnants of bacteria.

Methanogenic Digestate

Methanogenic digestate is a sludge (sometimes called a liquor). This is often high in nutrients such as ammoniums and phosphates.

Uses

The primary use of digestate is as a soil conditioner. Acidogenic digestate provides moisture retention and organic content for soils. This organic material can break down further, aerobically in soil. Methanogenic digestate provides nutrients for plant growth. It can also be used to protect soils against erosion.

Acidogenic digestate can also be used as an environmentally friendly filler to give structure to composite plastics.

Growth trials on digestate originating from mixed waste have showed healthy growth results for crops.

Application of digestate has been shown to inhibit plant diseases and induction of resistance. Digestate application has a direct effect on soil-born diseases, and an indirect effect by stimulation of biological activity.

Digestate and Compost

Digestate is technically not compost although it is similar to it in physical and chemical characteristics. Compost is produced by aerobic digestion- decomposition by aerobes. This includes fungi and bacteria which are able to break down the lignin and cellulose to a greater extent.

Standards for Digestate

The standard of digestate produced by anaerobic digestion can be assessed on three criteria, chemical, biological and physical aspects. Chemical quality needs to be considered in terms of heavy metals and other inorganic contaminant, persistent organic compounds and the content of macro-elements such as Nitrogen, Phosphourous and Potassium. Depending on their source, biowastes can contain pathogens, which can lead to the spreading of human, animal or plant diseases if not appropriately managed.

The physical standards of composts includes mainly appearance and odour factors. Whilst physical contamination does not present a problem with regards to human, plant or animal health, contamination (in the form of plastics, metals and ceramics) can cause a negative public perception. Even if the compost is of high quality and all standards are met, a negative public perception of waste-based composts still exists. The presence of visible contaminants reminds users of this.

Quality control of the feedstock is the most important way of ensuring a quality end product. The content and quality of waste arriving on-site should be characterised as thoroughly as possible prior to being supplied.

In the UK the Publicly Available Specification (called PAS110) governs the definition of digestate derived from the anaerobic digestion of source-segregated biodegradable materials. The specification ensures all digested materials are of consistent quality and fit for purpose. If a biogas plant meets the standard, its digestate will be regarded as having been fully recovered and to have ceased to be waste, and it can be sold with the name "Bio- fertiliser".

Source Separated Organics

Source Separated Organics (SSO) is the system by which waste generators segregate compostable materials from other waste streams at the source for separate collection.

A resident adds kitchen food scraps to yard debris in a roll cart as part of the community's source separated organics (SSO) program.

Types of Materials

Organic materials, such as yard trimmings, food scraps, wood waste, and paper and paperboard products, typically make up about one-third (by weight) of the municipal solid waste stream. SSO programs depend on the composition of local waste stream, acceptance specifications for the organics processing facility, and collection methods. The types of organic materials collected include:

- Yard and landscaping debris: floral trimmings, tree trimmings, leaves, grass, brush, and weeds

- Food waste: organic residues generated by the handling, storage, sale, preparation, cooking, and serving of foods, including fruits, vegetables, meat, poultry, seafood, shellfish, bones, rice, beans, pasta, bakery items, cheese, eggshells, and coffee grounds

- Paper fibers: waxed cardboard, napkins, paper towels, uncoated paper plates, tea bags, coffee filters, wooden crates, and greasy pizza boxes

- Wood waste: urban wood waste, woody debris from suburban land clearing, and rural forestry residuals

Programs

SSO programs have been launched in a wide range of venues, including single-family residential units, commercial businesses, events, food processors, schools, hospitals, and airports. The U.S. Environmental Protection Agency (EPA) has assembled tools and resources for food waste management to assist communities interested in launching their own food waste reduction and collection efforts. SSO materials are typically collected in wet-strength paper bags, unlined plastic bins, or compostable film-plastic liners that meet ASTM 6400 standards.

Benefits

The organic fraction of the waste stream is increasingly viewed as a resource. The resulting products – renewable energy and compost – benefit the environment: reduce greenhouse gas emissions; reduce dependency on foreign energy imports; increase the nutrient composition of our soils; reduce the amount of waste going to landfills; reduce the amount of wet, sloppy waste going to other methods of disposal; reduce the leachate associated with stormwater management at landfills; reduce the greenhouse gas emissions from uncontrolled landfill operations; improve erosion and stormwater control through biofiltration (Schwab, 2000).

Barriers to Adoption

Communities and businesses that want to implement SSO programs face a few challenges. First, they need participation at the source of their organic waste generation. Second, they need a hauler willing to collect the organic waste. Third, they need a composting facility permitted to process the material. These challenges have been overcome by many successful SSO programs. Tactics for addressing barriers to adoption include creating outreach and education materials, forging partnerships between local businesses to share fixed collection costs, and creating incentives for organic diversion through regulated tip fees for solid waste and organics.

Processing

Organic materials collected in SSO programs typically get delivered to composting facilities where the waste is turned into nutrient-rich soil amendments known as compost. Organic feedstock can also be delivered to anaerobic digestion facilities that produce biogas, a source of renewable energy. Anaerobic digestion of the organic fraction of MSW Municipal Solid Waste has been found to be in a number of LCA analysis studies to be more environmentally effective, than landfill, incineration or pyrolysis. The resulting biogas (methane) can then be used for cogeneration (electricity and heat preferably on or close to the site of production) and can be used in gas combustion engines or turbines. With further upgrading to synthetic natural gas it can be injected into the natural gas network or further refined to hydrogen for use in stationary cogeneration fuel cells.

References

- United Nations Environmental Programme (2013). "Guidelines for National Waste Management Strategies Moving from Challenges to Opportunities." (PDF). ISBN 978-92-807-3333-4.

- Gandy, Matthew (1994). Recycling and the Politics of Urban Waste. Earthscan. ISBN 9781853831683.

- Villadsen, John; Nielsen, Jens Høiriis; Lidén, Gunnar (2003). Bioreaction Engineering Principles (2 ed.). Springer. ISBN 978-0-306-47349-4.

- Spakowicz, Daniel J.; Strobel, Scott A. (2015). "Biosynthesis of hydrocarbons and volatile organic compounds by fungi: bioengineering potential". Applied microbiology and biotechnology. 99 (12): 4943–4951. Retrieved 2016-02-22.

- The National Academies Press (2008). "Water Issues of Biofuel Production Plants". The National Academies Press. Retrieved 18 June 2015.

- REN21 (2011). "Renewables 2011: Global Status Report" (PDF). pp. 13–14. Archived from the original (PDF) on 2011-09-05. Retrieved 2015-01-03.

- Ramirez, Jerome; Brown, Richard; Rainey, Thomas (1 July 2015). "A Review of Hydrothermal Liquefaction Bio-Crude Properties and Prospects for Upgrading to Transportation Fuels". Energies. 8: 6765–6794. doi:10.3390/en8076765.

- "The potential and challenges of drop-in fuels (members only) | IEA Bioenergy Task 39 – Commercializing Liquid Biofuels". task39.sites.olt.ubc.ca. Retrieved 2015-09-10.

- "Sweet Sorghum : A New "Smart Biofuel Crop"". Agriculture Business Week. 30 June 2008. Archived from the original on 27 May 2015.

- Runge, Ford, and Benjamin Senauer. "How Biofuels Could Starve the Poor" Foreign Affairs 86 (2007): 41-53. Accessed October 30, 2014. from: http://www.jstor.org/stable/20032348

- Rock, Kerry; Maurice Korpelshoek (2007). [<http://www.digitalrefining.com/article_1000210 "Bioethers Impact on the Gasoline Pool"] Check |url= value (help). Digital Refining. Retrieved 15 February 2014.

- Sukla, Mirtunjay Kumar; Thallada Bhaskar; A.K. Jain; S.K. Singal; M.O. Garg. "Bio-Ethers as Transportation Fuel: A Review"] Check |url= value (help) (PDF). Indian Institute of Petroleum Dehradun. Retrieved 15 February 2014.

- Waste Management (2013). "Editorial Board/Aims & Scopes". Waste Management. 34: IFC. doi:10.1016/S0956-053X(14)00026-9.

- Science Direct (2013). "Waste Management". Volume 33, Issue 1 pp220-232.

- Bogorad, I. W.; Lin, T. S.; Liao, J. C. (2013). "Synthetic non-oxidative glycolysis enables complete carbon conservation". Nature. doi:10.1038/nature12575.

- Choi, Y. J.; Lee, S. Y. (2013). "Microbial production of short-chain alkanes". Nature. 502: 571–4. doi:10.1038/nature12536. PMID 24077097.

- Ethanol Research (2012-04-02). "National Corn-to-Ethanol Research Center (NCERC)". Ethanol Research. Archived from the original on 20 March 2012. Retrieved 2012-04-02.

- National Renewable Energy Laboratory (2007-03-02). "Research Advantages: Cellulosic Ethanol" (PDF). National Renewable Energy Laboratory. Retrieved 2012-04-02.

- Sheehan, John; et al. (July 1998). "A Look Back at the U. S. Department of Energy's Aquatic Species Program: Biofuels from Algae" (PDF). National Renewable Energy Laboratory. Retrieved 16 June 2012.

Methods and Techniques of Biodegradabale Waste Management

Methods and techniques are an important component of any field of study. The following chapter elucidates the various techniques that are related to biodegradable waste management. Promession, bioremediation, pyrolysis, rotating biological contactor are some of the methods that are covered in this section.

Promession

Promession is an environmentally friendly way to dispose of human remains by way of freeze drying. The concept of promession was developed by Swedish biologist Susanne Wiigh-Mäsak, who derived the name from the Italian word for "promise" (*promessa*). She founded Promessa Organic AB in 1997 to commercially pursue her idea.

Process

Promession involves five steps:

1. Coffin separation: the body is placed into the chamber

2. Cryogenic freezing: liquid nitrogen at -196 °C crystallizes the body

3. Vibration: the body is disintegrated into particles within minutes

4. Freeze drying: particles are freeze dried in a drying chamber, leaving approximately 30% of the original weight

5. Metal separation: any metals (e.g., tooth amalgam, artificial hips, etc.) are removed, either by magnetism or by sieving. The dry powder is placed in a biodegradable casket which is interred in the top layers of soil, where aerobic bacteria decompose the remains into humus in as little as 6–12 months.

Current Status

From 2004, trials have been performed on pigs, and AGA Gas developed a proof-of-concept. However a third party is needed to enter into an agreement with Promessa to order the equipment needed for promession of human cadavers.

The BBC has shown a proof of concept to work with relatively simple means.

Wiigh-Mäsak had received expressions of interest from more than 60 countries, including Vietnam, the United Kingdom, South Africa, the Netherlands, Canada, and the United States. In South Korea, the technology was expressly legalized. Currently, Wiigh-Mäsak works with groups, countries, and people of all kinds to find support for her company and lifelong passion, encouraging others to show support through membership and donation for Promessa.

Public Opinion

An opinion poll run by Ny Teknik in Sweden showed support for promession. In a popularity contest among about 70 innovative companies in Sweden, Promessa was judged the most popular.

Bioremediation

Bioremediation is a waste management technique that involves the use of organisms to remove or neutralize pollutants from a contaminated site. According to the United States EPA, bioremediation is a "treatment that uses naturally occurring organisms to break down hazardous substances into less toxic or non toxic substances". Technologies can be generally classified as *in situ* or *ex situ*. *In situ* bioremediation involves treating the contaminated material at the site, while *ex situ* involves the removal of the contaminated material to be treated elsewhere. Some examples of bioremediation related technologies are phytoremediation, bioventing, bioleaching, landfarming, bioreactor, composting, bioaugmentation, rhizofiltration, and biostimulation.

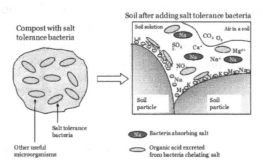

Mechanism of salt removal from tsunami affected soil by bioremediation

Bioremediation may occur on its own (natural attenuation or intrinsic bioremediation) or may only effectively occur through the addition of fertilizers, oxygen, etc.,that help in enhancing the growth of the pollution-eating microbes within the medium (biostimulation). For example, the US Army Corps of Engineers demonstrated that windrowing and aeration of petroleum-contaminated soils enhanced bioremediation using the technique of landfarming. Depleted soil nitrogen status may encourage biodegradation of some nitrogenous organic chemicals, and soil materials with a high capacity to adsorb pollutants may slow down biodegradation owing to limited bioavailability of the chemicals to microbes. Recent advancements have also proven successful via the addition of matched microbe strains to the medium to enhance the resident microbe population's ability to break down contaminants. Microorganisms used to perform the function of bioremediation are known as bioremediators.

However, not all contaminants are easily treated by bioremediation using microorganisms. For example, heavy metals such as cadmium and lead are not readily absorbed or captured by microorganisms. A recent experiment, however, suggests that fish bones have some success absorbing lead from contaminated soil. Bone char has been shown to bioremediate small amounts of cadmium, copper, and zinc. A recent experiment, suggests that the removals of pollutants (nitrate, silicate, chromium and sulphide) from tannery wastewater were studied in batch experiments using marine microalgae. The assimilation of metals such as mercury into the food chain may worsen matters. Phytoremediation is useful in these circumstances because natural plants or transgenic plants are able to bioaccumulate these toxins in their above-ground parts, which are then harvested for removal. The heavy metals in the harvested biomass may be further concentrated by incineration or even recycled for industrial use. Some damaged artifacts at museums contain microbes which could be specified as bio remediating agents. In contrast to this situation, other contaminants, such as aromatic hydrocarbons as are common in petroleum, are relatively simple targets for microbial degradation, and some soils may even have some capacity to autoremediate, as it were, owing to the presence of autochthonous microbial communities capable of degrading these compounds.

The elimination of a wide range of pollutants and wastes from the environment requires increasing our understanding of the relative importance of different pathways and regulatory networks to carbon flux in particular environments and for particular compounds, and they will certainly accelerate the development of bioremediation technologies and biotransformation processes.

Genetic Engineering Approaches

The use of genetic engineering to create organisms specifically designed for bioremediation has great potential. The bacterium *Deinococcus radiodurans* (the most radioresistant organism known) has been modified to consume and digest toluene and ionic mercury from highly radioactive nuclear waste. Releasing genetically augmented organisms into the environment may be problematic as tracking them can be difficult; bioluminescence genes from other species may be inserted to make this easier.

Mycoremediation

Mycoremediation is a form of bioremediation in which fungi are used to decontaminate the area. The term *mycoremediation* refers specifically to the use of fungal mycelia in bioremediation.

One of the primary roles of fungi in the ecosystem is decomposition, which is performed by the mycelium. The mycelium secretes extracellular enzymes and acids that break down lignin and cellulose, the two main building blocks of plant fiber. These are organic compounds composed of long chains of carbon and hydrogen, structurally similar to many organic pollutants. The key to mycoremediation is determining the right fungal species to target a specific pollutant. Certain strains have been reported to successfully degrade the nerve gases VX and sarin.

In one conducted experiment, a plot of soil contaminated with diesel oil was inoculated with mycelia of oyster mushrooms; traditional bioremediation techniques (bacteria) were used on control plots. After four weeks, more than 95% of many of the PAH (polycyclic aromatic hydrocarbons) had been reduced to non-toxic components in the mycelial-inoculated plots. It appears that the natural microbial community participates with the fungi to break down contaminants, eventually into carbon dioxide and water. Wood-degrading fungi are particularly effective in breaking down aromatic pollutants (toxic components of petroleum), as well as chlorinated compounds (certain persistent pesticides; Battelle, 2000).

Two species of the Ecuadorian fungus Pestalotiopsis are capable of consuming Polyurethane in aerobic and anaerobic conditions such as found at the bottom of landfills.

Mycofiltration is a similar process, using fungal mycelia to filter toxic waste and microorganisms from water in soil.

Advantages

There are a number of cost/efficiency advantages to bioremediation, which can be employed in areas that are inaccessible without excavation. For example, hydrocarbon spills (specifically, petrol spills) or certain chlorinated solvents may contaminate groundwater, and introducing the appropriate electron acceptor or electron donor amendment, as appropriate, may significantly reduce contaminant concentrations after a long time allowing for acclimation. This is typically much less expensive than excavation followed by disposal elsewhere, incineration or other *ex situ* treatment strategies, and reduces or eliminates the need for "pump and treat", a practice common at sites where hydrocarbons have contaminated clean groundwater. Using archaea for bioremediation of hydrocarbons also has the advantage of breaking down contaminants at the molecular level, as opposed to simply chemically dispersing the contaminant.

Monitoring Bioremediation

The process of bioremediation can be monitored indirectly by measuring the *Oxidation Reduction Potential* or redox in soil and groundwater, together with pH, temperature, oxygen content, electron acceptor/donor concentrations, and concentration of breakdown products (e.g. carbon dioxide). This table shows the (decreasing) biological breakdown rate as function of the redox potential.

Process	Reaction	Redox potential (E_h in mV)
aerobic	$O_2 + 4e^- + 4H^+ \rightarrow 2H_2O$	600 ~ 400
anaerobic		
denitrification	$2NO_3^- + 10e^- + 12H^+ \rightarrow N_2 + 6H_2O$	500 ~ 200
manganese IV reduction	$MnO_2 + 2e^- + 4H^+ \rightarrow Mn^{2+} + 2H_2O$	400 ~ 200
iron III reduction	$Fe(OH)_3 + e^- + 3H^+ \rightarrow Fe^{2+} + 3H_2O$	300 ~ 100
sulfate reduction	$SO_4^{2-} + 8e^- + 10 H^+ \rightarrow H_2S + 4H_2O$	0 ~ −150
fermentation	$2CH_2O \rightarrow CO_2 + CH_4$	−150 ~ −220

This, by itself and at a single site, gives little information about the process of remediation.

1. It is necessary to sample enough points on and around the contaminated site to be able to determine contours of equal redox potential. Contouring is usually done using specialised software, e.g. using Kriging interpolation.

2. If all the measurements of redox potential show that electron acceptors have been used up, it is in effect an indicator for total microbial activity. Chemical analysis is also required to determine when the levels of contaminants and their breakdown products have been reduced to below regulatory limits.

3. Chemical analysis should also be carried out for assessing transformations in inorganic contaminants (e.g. heavy metals, radionuclides). Unlike organic pollutants, inorganic pollutants cannot be degraded and remediation processes can both increase and decrease their solubility and bio-availability. An increase in heavy metal mobility can occur, even in reductive conditions, during *in-situ* bioremediation.

Pyrolysis

Pyrolysis is a thermochemical decomposition of organic material at elevated temperatures in the absence of oxygen (or any halogen). It involves the simultaneous change of chemical composition and physical phase, and is irreversible. The word is coined from the Greek-derived elements *pyro* "fire" and *lysis* "separating".

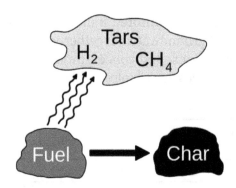

Simplified depiction of pyrolysis chemistry.

Pyrolysis is a type of thermolysis, and is most commonly observed in organic materials exposed to high temperatures. It is one of the processes involved in charring wood, starting at 200–300 °C (390–570 °F). It also occurs in fires where solid fuels are burning or when vegetation comes into contact with lava in volcanic eruptions. In general, pyrolysis of organic substances produces gas and liquid products and leaves a solid residue richer in carbon content, char. Extreme pyrolysis, which leaves mostly carbon as the residue, is called carbonization.

The process is used heavily in the chemical industry, for example, to produce charcoal, activated carbon, methanol, and other chemicals from wood, to convert ethylene dichloride into vinyl chloride to make PVC, to produce coke from coal, to convert biomass into syngas and biochar, to turn waste plastics back into usable oil, or waste into safely disposable substances, and for transforming medium-weight hydrocarbons from oil into lighter ones like gasoline. These specialized uses of pyrolysis may be called various names, such as dry distillation, destructive distillation, or cracking. Pyrolysis is also used in the creation of nanoparticles, zirconia and oxides utilizing an ultrasonic nozzle in a process called ultrasonic spray pyrolysis (USP).

Pyrolysis also plays an important role in several cooking procedures, such as baking, frying, grilling, and caramelizing. It is a tool of chemical analysis, for example, in mass spectrometry and in carbon-14 dating. Indeed, many important chemical substances, such as phosphorus and sulfuric acid, were first obtained by this process. Pyrolysis has been assumed to take place during catagenesis, the conversion of buried organic matter to fossil fuels. It is also the basis of pyrography. In their embalming process, the ancient Egyptians used a mixture of substances, including methanol, which they obtained from the pyrolysis of wood.

Pyrolysis differs from other processes like combustion and hydrolysis in that it usually does not involve reactions with oxygen, water, or any other reagents. In practice, it is not possible to achieve a completely oxygen-free atmosphere. Because some oxygen is present in any pyrolysis system, a small amount of oxidation occurs.

The term has also been applied to the decomposition of organic material in the presence of super-heated water or steam (hydrous pyrolysis), for example, in the steam cracking of oil.

Occurrence and Uses
Fire

Pyrolysis is usually the first chemical reaction that occurs in the burning of many solid organic fuels, like wood, cloth, and paper, and also of some kinds of plastic. In a wood fire, the visible flames are not due to combustion of the wood itself, but rather of the gases released by its pyrolysis, whereas the flame-less burning of a solid, called smouldering, is the combustion of the solid residue (char or charcoal) left behind by pyrolysis. Thus, the pyrolysis of common materials like wood, plastic, and clothing is extremely important for fire safety and firefighting. In pyrolysis there is a gas phase present. It should not be confused with hydrothermal reactions such as hydrothermal gasification, hydrothermal liquidation, and hydrothermal carbonization, which occur in aqueous environments because the temperatures and reaction pathways differ, with ionic reactions favored in aqueous reactions and radical reactions favored in the absence of water.

Cooking

Pyrolysis occurs whenever food is exposed to high enough temperatures in a dry environment, such as roasting, baking, toasting, or grilling. It is the chemical process responsible for the formation of the golden-brown crust in foods prepared by those methods.

In normal cooking, the main food components that undergo pyrolysis are carbohydrates (including sugars, starch, and fibre) and proteins. Pyrolysis of fats requires a much higher temperature, and, since it produces toxic and flammable products (such as acrolein), it is, in general, avoided in normal cooking. It may occur, however, when grilling fatty meats over hot coals.

Even though cooking is normally carried out in air, the temperatures and environmental conditions are such that there is little or no combustion of the original substances or their decomposition products. In particular, the pyrolysis of proteins and carbohydrates begins at temperatures much lower than the ignition temperature of the solid residue, and the volatile subproducts are too diluted in air to ignite. (In flambé dishes, the flame is due mostly to combustion of the alcohol, while the crust is formed by pyrolysis as in baking.)

Pyrolysis of carbohydrates and proteins requires temperatures substantially higher than 100 °C (212 °F), so pyrolysis does not occur as long as free water is present, e.g., in boiling food — not even in a pressure cooker. When heated in the presence of water, carbohydrates and proteins suffer gradual hydrolysis rather than pyrolysis. Indeed, for most foods, pyrolysis is usually confined to the outer layers of food, and begins only after those layers have dried out.

Food pyrolysis temperatures are, however, lower than the boiling point of lipids, so pyrolysis occurs when frying in vegetable oil or suet, or basting meat in its own fat.

Pyrolysis also plays an essential role in the production of barley tea, coffee, and roasted nuts such as peanuts and almonds. As these consist mostly of dry materials, the process of pyrolysis is not limited to the outermost layers but extends throughout the materials. In all these cases, pyrolysis creates or releases many of the substances that contribute to the flavor, color, and biological properties of the final product. It may also destroy some substances that are toxic, unpleasant in taste, or those that may contribute to spoilage.

Controlled pyrolysis of sugars starting at 170 °C (338 °F) produces caramel, a beige to brown water-soluble product widely used in confectionery and (in the form of caramel coloring) as a coloring agent for soft drinks and other industrialized food products.

Solid residue from the pyrolysis of spilled and splattered food creates the brown-black encrustation often seen on cooking vessels, stove tops, and the interior surfaces of ovens.

Charcoal

Pyrolysis has been used since ancient times for turning wood into charcoal on an industrial scale. Besides wood, the process can also use sawdust and other wood waste products.

Charcoal is obtained by heating wood until its complete pyrolysis (carbonization) occurs, leaving only carbon and inorganic ash. In many parts of the world, charcoal is still produced semi-industrially, by burning a pile of wood that has been mostly covered with mud or bricks. The heat generated by burning part of the wood and the volatile byproducts pyrolyzes the rest of the pile. The limited supply of oxygen prevents the charcoal from burning. A more modern alternative is to heat the wood in an airtight metal vessel, which is much less polluting and allows the volatile products to be condensed.

The original vascular structure of the wood and the pores created by escaping gases combine to produce a light and porous material. By starting with a dense wood-like material, such as nutshells or peach stones, one obtains a form of charcoal with particularly fine pores (and hence a much larger pore surface area), called activated carbon, which is used as an adsorbent for a wide range of chemical substances.

Biochar

Residues of incomplete organic pyrolysis, e.g., from cooking fires, are thought to be the key component of the terra preta soils associated with ancient indigenous communities of the Amazon basin. Terra preta is much sought by local farmers for its superior fertility compared to the natural red soil of the region. Efforts are underway to recreate these soils through biochar, the solid residue of pyrolysis of various materials, mostly organic waste.

Biochar improves the soil texture and ecology, increasing its ability to retain fertilizers and release them slowly. It naturally contains many of the micronutrients needed by plants, such as selenium. It is also safer than other "natural" fertilizers such as animal manure, since it has been disinfected at high temperature. And, since it releases its nutrients at a slow rate, it greatly reduces the risk of water table contamination.

Biochar is also being considered for carbon sequestration, with the aim of mitigation of global warming. The solid, carbon-containing char produced can be sequestered in the ground, where it will remain for several hundred to a few thousand years.

Coke

Pyrolysis is used on a massive scale to turn coal into coke for metallurgy, especially steelmaking. Coke can also be produced from the solid residue left from petroleum refining.

Those starting materials typically contain hydrogen, nitrogen, or oxygen atoms combined with carbon into molecules of medium to high molecular weight. The coke-making or "coking" process consists of heating the material in closed vessels to very high temperatures (up to 2,000 °C or 3,600 °F) so that those molecules are broken down into lighter volatile substances, which leave the vessel, and a porous but hard residue that is mostly carbon and inorganic ash. The amount of volatiles varies with the source material, but is typically 25–30% of it by weight.

Carbon Fiber

Carbon fibers are filaments of carbon that can be used to make very strong yarns and textiles. Carbon fiber items are often produced by spinning and weaving the desired item from fibers of a suitable polymer, and then pyrolyzing the material at a high temperature (from 1,500–3,000 °C or 2,730–5,430 °F).

The first carbon fibers were made from rayon, but polyacrylonitrile has become the most common starting material.

For their first workable electric lamps, Joseph Wilson Swan and Thomas Edison used carbon filaments made by pyrolysis of cotton yarns and bamboo splinters, respectively.

Pyrolytic Carbon

Pyrolysis is the reaction used to coat a preformed substrate with a layer of pyrolytic carbon. This is typically done in a fluidized bed reactor heated to 1,000–2,000 °C or 1,830–3,630 °F. Pyrolytic carbon coatings are used in many applications, including artificial heart valves.

Biofuel

Pyrolysis is the basis of several methods that are being developed for producing fuel from biomass, which may include either crops grown for the purpose or biological waste products from other industries. Crops studied as biomass feedstock for pyrolysis include native North American prairie grasses such as *switchgrass* and bred versions of other grasses such as *Miscantheus giganteus*. Crops and plant material wastes provide biomass feedstock on the basis of their lignocellulose portions.

Although synthetic diesel fuel cannot yet be produced directly by pyrolysis of organic materials, there is a way to produce similar liquid (bio-oil) that can be used as a fuel, after the removal of valuable bio-chemicals that can be used as food additives or pharmaceuticals. Higher efficiency is achieved by the so-called flash pyrolysis, in which finely divided feedstock is quickly heated to between 350 and 500 °C (660 and 930 °F) for less than 2 seconds.

Fuel bio-oil can also be produced by hydrous pyrolysis from many kinds of feedstock, including waste from pig and turkey farming, by a process called thermal depolymerization (which may, however, include other reactions besides pyrolysis).

Adhesives

Neanderthals used pyrolysis of birch bark to produce a pitch with which they secured flaked stones to spear shafts. Recently, researchers have developed a process to pyrolyze birch bark to produce an oil that can replace phenol in phenol formaldehyde resin (these resins are mostly used to manufacture plywood).

Pesticides

Pyrolysis can also be used to produce pesticides from biomass.

Plastic Waste Disposal

Anhydrous pyrolysis can also be used to produce liquid fuel similar to diesel from plastic waste, with a higher cetane value and lower sulphur content than traditional diesel. Using pyrolysis to extract fuel from end-of-life plastic is a second-best option after recycling, is environmentally preferable to landfill, and can help reduce dependency on foreign fossil fuels and geo-extraction. Pilot Jeremy Roswell plans to make the first flight from Sydney to London using diesel fuel from recycled plastic waste manufactured by Cynar PLC.

Waste Tire Disposal

In the United States alone, over 290 million car tires are discarded annually. Pyrolysis of scrap or waste tires (WT) is an attractive alternative to disposal in landfills, allowing the high energy content of the tire to be recovered as fuel. Using tires as fuel produce equal energy as burning oil and 25% more energy than burning coal.

An average car tire is made up of 50-60% hydrocarbons, resulting in a yield of 38-56% oil, 10-30% gas and 14-56% char. The oil produced is largely composed of benzene, diesel, kerosene, fuel oil and heavy fuel oil, while the produced gas has a similar composition to natural gas. The proportion and the purity of the products are governed by two major factors:

1. Environment (e.g. pressure, temperature, time, reactor type)

2. Material (e.g. age, composition, size, type)

As car tires age, they increase in hardness, making it more difficult for pyrolysis to break the molecules into shorter chains. This shifts the yield composition towards diesel oil which is composed of larger molecules. Conversely, an increase in temperature increases the likelihood of breaking the molecule chain and shifts the yield composition towards benzene oil which is composed of smaller molecules. Other products from car tire pyrolysis include steel wires, carbon black and bitumen.

Although the pyrolysis of WT has been widely developed throughout the world, there are legislative, economic, and marketing obstacles to widespread adoption. Oil derived from tire rubber pyrolysis contains high sulfur content, which gives it high potential as a pollutant and should be desulfurized A number of prototype and full-scale pyrolysis plants specialized in carbon black production have successfully established across the world, including the United States, France, Germany and Japan. Because carbon black is used for pigment, rubber strengthening and UV protection, it is a relatively large and growing market. Pyrolysis plants specialized in fuel oil production is not an implausible concept. However, as profits of such ventures come from the added value between the production and distillation of oil, there is little profit without vertical integration in the oil industry. The inconsistency of the feedstock makes it very difficult to control the uniformity of the products and makes oil companies hesitant to purchase oil produced via pyrolysis. Finally, the cost of producing oil through conventional means is generally less expensive than this alternative. To date, there is no known commercially profitable standalone pyrolysis plant that specializes in oil production. However, with funding to upgrade pyrolysis oil to light fuel grade, this may be possible. Nevertheless, pyrolysis is a valuable method for disposing waste tires.

Chemical Analysis

Pyrolysis can be used for the molecular characterisation of molecules when used in conjunction with gas chromatography-mass spectrometry (Py-GC-MS). This technique has been used to analyse the method and products of fungal decay of wood.

Thermal Cleaning

Pyrolysis is also used for *thermal cleaning*, an industrial application to remove organic substances such as polymers, plastics and coatings from parts, products or production components like extruder screws, spinnerets and static mixers. During the thermal cleaning process, at temperatures between 600 °F to 1000 °F (310 C° to 540 C°), organic material is converted by pyrolysis and oxidation into volatile organic compounds, hydrocarbons and carbonized gas. Inorganic elements remain.

Several types of thermal cleaning systems use pyrolysis:

- *Molten Salt Baths* belong to the oldest thermal cleaning systems; cleaning with a molten salt bath is very fast but implies the risk of dangerous splatters, or other potential hazards connected with the use of salt baths, like explosions or highly toxic hydrogen cyanide gas;

- *Fluidized Bed Systems* use sand or aluminium oxide as heating medium; these systems also clean very fast but the medium does not melt or boil, nor emit any vapors or odors; the cleaning process takes one to two hours;

- *Vacuum Ovens* use pyrolysis in a vacuum avoiding uncontrolled combustion inside the cleaning chamber; the cleaning process takes 8 to 30 hours;

- *Burn-Off Ovens*, also known as *Heat-Cleaning Ovens*, are gas-fired and used in the painting, coatings, electric motors and plastics industries for removing organics from heavy and large metal parts.

Processes

In many industrial applications, the process is done under pressure and at operating temperatures above 430 °C (806 °F). For agricultural waste, for example, typical temperatures are 450 to 550 °C (840 to 1,000 °F).

Processes

Since pyrolysis is endothermic, various methods to provide heat to the reacting biomass particles have been proposed:

- Partial combustion of the biomass products through air injection. This results in poor-quality products.

- Direct heat transfer with a hot gas, the ideal one being product gas that is reheated and recycled. The problem is to provide enough heat with reasonable gas flow-rates.

- Indirect heat transfer with exchange surfaces (wall, tubes): it is difficult to achieve good heat transfer on both sides of the heat exchange surface.

- Direct heat transfer with circulating solids: solids transfer heat between a burner and a pyrolysis reactor. This is an effective but complex technology.

For flash pyrolysis, the biomass must be ground into fine particles and the insulating char layer that forms at the surface of the reacting particles must be continuously removed. The following technologies have been proposed for biomass pyrolysis:

- Fixed beds used for the traditional production of charcoal: poor, slow heat transfer result in very low liquid yields.

- Augers: this technology is adapted from a Lurgi process for coal gasification. Hot sand and biomass particles are fed at one end of a screw. The screw mixes the sand and biomass and conveys them along. It provides a good control of the biomass residence time. It does not dilute the pyrolysis products with a carrier or fluidizing gas. However, sand must be reheated in a separate vessel, and mechanical reliability is a concern. There is no large-scale commercial implementation.

- Electrically heated augers: one process uses an electrical current passed through an auger to heat the material giving excellent heat transfer by contact and radiation to the waste material.

- Ablative processes: biomass particles are moved at high speed against a hot metal surface. Ablation of any char forming at a particle's surface maintains a high rate of heat transfer. This can be achieved by using a metal surface spinning at high speed within a bed of biomass particles, which may present mechanical reliability problems but prevents any dilution of the products. As an alternative, the particles may be suspended in a carrier gas and introduced at high speed through a cyclone whose wall is heated; the products are diluted with the carrier gas. A problem shared with all ablative processes is that scale-up is made difficult, since the ratio of the wall surface to the reactor volume decreases as the reactor size is increased. There is no large-scale commercial implementation.

- Rotating cone: pre-heated hot sand and biomass particles are introduced into a rotating cone. Due to the rotation of the cone, the mixture of sand and biomass is transported across the cone surface by centrifugal force. The process is offered by BTG-BTL, a subsidiary from BTG Biomass Technology Group B.V. in The Netherlands. Like other shallow transported-bed reactors relatively fine particles (several mm) are required to obtain a liquid yield of around 70 wt.%. Larger-scale commercial implementation (up to 5 t/h input) is underway.

- Fluidized beds: biomass particles are introduced into a bed of hot sand fluidized by a gas, which is usually a recirculated product gas. High heat transfer rates from fluidized sand result in rapid heating of biomass particles. There is some ablation by attrition with the sand particles, but it is not as effective as in the ablative processes. Heat is usually provided by heat exchanger tubes through which hot combustion gas flows. There is some dilution of the products, which makes it more difficult to condense and then remove the bio-oil mist from the gas exiting the condensers. This process has been scaled up by companies such as Dynamotive and Agri-Therm. The main challenges are in improving the quality and consistency of the bio-oil.

- Circulating fluidized beds: biomass particles are introduced into a circulating fluidized bed of hot sand. Gas, sand, and biomass particles move together, with the transport gas usually being a recirculated product gas, although it may also be a combustion gas. High heat transfer rates from sand ensure rapid heating of biomass particles and ablation stronger than with regular fluidized beds. A fast separator separates the product gases and vapors from the sand and char particles. The sand particles are reheated in a fluidized burner vessel and recycled to the reactor. Although this process can be easily scaled up, it is rather complex and the products are much diluted, which greatly complicates the recovery of the liquid products.

- Mechanical Fluidized Reactor (MFR). A mechanical stirrer agitates a hot bed of pure char particles into which biomass particles are injected. The stirrer also enhances heat transfer from the reactor wall to the agitated bed. No fluidization gas is required: evolving vapors aerate the bed and greatly reduce the power consumption of the mechanical stirrer. This compact reactor has been used for a mobile pyrolysis plant.

- Chain grate: dry biomass is fed onto a hot (500 °C) heavy cast metal grate or apron which forms a continuous loop. A small amount of air aids in heat transfer and in primary reactions for drying and carbonization. Volatile products are combusted for process and boiler heating.

Use of Vacuum

In vacuum pyrolysis, organic material is heated in a vacuum to decrease its boiling point and avoid adverse chemical reactions. Called flash vacuum pyrolysis, this approach is used in organic synthesis.

Industrial Sources

Many sources of organic matter can be used as feedstock for pyrolysis. Suitable plant material includes greenwaste, sawdust, waste wood, woody weeds; and agricultural sources including nut shells, straw, cotton trash, rice hulls, switch grass; and animal waste including poultry litter, dairy manure, and potentially other manures. Pyrolysis is used as a form of thermal treatment to reduce waste volumes of domestic refuse. Some industrial byproducts are also suitable feedstock including paper sludge and distillers grain.

There is also the possibility of integrating with other processes such as mechanical biological treatment and anaerobic digestion.

Industrial Products

- syngas (flammable mixture of carbon monoxide and hydrogen): can be produced in sufficient quantities to provide both the energy needed for pyrolysis and some excess production
- solid char that can either be burned for energy or be recycled as a fertilizer (biochar).

Fire Protection

Destructive fires in buildings will often burn with limited oxygen supply, resulting in pyrolysis reactions. Thus, pyrolysis reaction mechanisms and the pyrolysis properties of materials are important in fire protection engineering for passive fire protection. Pyrolytic carbon is also important to fire investigators as a tool for discovering origin and cause of fires.

Chemistry

Current research examines the multiple reaction pathways of pyrolysis to understand how to manipulate the formation of pyrolysis' multiple products (oil, gas, char, and miscellaneous chemicals) to enhance the economic value of pyrolysis; identifying catalysts to manipulate pyrolysis reactions is also a goal of some pyrolysis research. Published research suggests that pyrolysis reactions have some dependence upon the structural composition of feedstocks (e.g. lignocellulosic biomass), with contributions from some minerals present in the feedstocks; some minerals present in feedstock are thought to increase the cost of operation of pyrolysis or decrease the value of oil produced from pyrolysis, through corrosive reactions. The low quality of oils produced through pyrolysis can be improved by subjecting the oils to one or many physical and chemical processes, which might drive production costs, but may make sense economically as circumstances change.

Rotating Biological Contactor

A rotating biological contactor or RBC is a biological treatment process used in the treatment of wastewater following primary treatment. The primary treatment process removes the grit and other solids through a screening process followed by a period of settlement. The RBC process involves allowing the wastewater to come in contact with a biological medium in order to remove pollutants in the wastewater before discharge of the treated wastewater to the environment, usually a body of water (river, lake or ocean). A rotating biological contactor is a type of secondary treatment process. It consists of a series of closely spaced, parallel discs mounted on a rotating shaft which is supported just above the surface of the waste water. Microorganisms grow on the surface of the discs where biological degradation of the wastewater pollutants takes place.

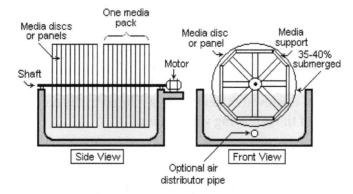

Schematic diagram of a typical rotating biological contactor (RBC). The treated effluent clarifier/settler is not included in the diagram.

Operation

The rotating packs of disks (known as the media) are contained in a tank or trough and rotate at between 2 and 5 revolutions per minute. Commonly used plastics for the media are polyethylene, PVC and expanded polystyrene. The shaft is aligned with the flow of wastewater so that the discs rotate at right angles to the flow with several packs usually combined to make up a treatment train. About 40% of the disc area is immersed in the wastewater.

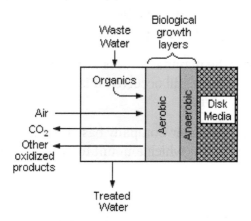

A schematic cross-section of the contact face of the bed media in a rotating biological contactor (RBC)

Biological growth is attached to the surface of the disc and forms a slime layer. The discs contact the wastewater with the atmospheric air for oxidation as it rotates. The rotation helps to slough off excess solids. The disc system can be staged in series to obtain nearly any detention time or degree of removal required. Since the systems are staged, the culture of the later stages can be acclimated to the slowly degraded materials.

The discs consist of plastic sheets ranging from 2 to 4 m in diameter and are up to 10 mm thick. Several modules may be arranged in parallel and/or in series to meet the flow and treatment requirements. The discs are submerged in waste water to about 40% of their diameter. Approximately 95% of the surface area is thus alternately submerged in waste water and then exposed to the atmosphere above the liquid. Carbonaceous substrate is removed in the initial stage of RBC. Carbon conversion may be completed in the first stage of a series of modules, with nitrification being completed after the 5th stage. Most design of RBC systems will include a minimum of 4 or 5 modules in series to obtain nitrification of waste water.

Biofilms, which are biological growths that become attached to the discs, assimilate the organic materials in the wastewater. Aeration is provided by the rotating action, which exposes the media to the air after contacting them with the wastewater, facilitating the degradation of the pollutants being removed. The degree of wastewater treatment is related to the amount of media surface area and the quality and volume of the inflowing wastewater.

History

The first RBC was installed in West Germany in 1960, later it was introduced in the United States and Canada. In the United States, rotating biological contactors are used for industries producing wastewaters high in biochemical oxygen demand (BOD) (e.g., petroleum industry and dairy industry).

A properly designed RBC can produce a very high quality final effluent. However both the organic and hydraulic loading have to be addressed. They do however suffer from low cycle fatigue and dependence on metal content, and designs suffer from short life failure.

Problems were encountered in the USA prompting the Environmental Agency to commission a number of reports.

These reports identified a number of issues and criticized the RBC process. One author suggested that since manufacturers were aware of the problem, the problems would be resolved and suggested that design engineers should specify a long life.

However, this is only possible if the manufacturer is aware of the design problems and the stress to ensure a long life and since failures still occurred it is unlikely any design stresses were widely published assuming they were known.

Severn Trent Water Ltd a large UK Water Company based in the Midlands employed these as the preferred process for their small works which amount to over 700 sites. Consequently, long life is essential to compliance.

This issue was successfully addressed by Eric Findlay C Eng when he was employed by Severn Trent Water Ltd in the UK following a period of failure of a number of plants. As a result, the issue of short life failure is now fully understood and is in the public domain and the correct process and hydraulic issues have been identified to produce a high quality nitrified effluent.

Since suppliers are responsible for the mechanical design of their plants purchasers should check that the system put in place is fully compliant with the procedures put in place by Findlay Cranfield University to avoid the risk of short life failure.

There are several papers by Findlay, Bannister and Mba which address this issue.

Secondary Clarification

Secondary clarifiers following RBCs are identical in design to conventional humus tanks, as used downstream of trickling filters. Sludge is generally removed daily, or pumped automatically to the primary settlement tank for co-settlement. Regular sludge removal reduces the risk of anaerobic conditions from developing within the sludge, with subsequent sludge flotation due to the release of gases.

Vermicompost

Vermicompost is the product or process of composting using various worms, usually red wigglers, white worms, and other earthworms, to create a heterogeneous mixture of decomposing vegetable or food waste, bedding materials, and vermicast, also called worm castings, worm humus or worm manure, is the end-product of the breakdown of organic matter by an earthworm. These castings have been shown to contain reduced levels of contaminants and a higher saturation of nutrients than do organic materials before vermicomposting.

Rotary screen harvested vermicompost, composed of worm castings

Containing water-soluble nutrients, vermicompost is an excellent, nutrient-rich organic fertilizer and soil conditioner. This process of producing vermicompost is called *vermicomposting*.

While vermicomposting is generally known as a nutrient rich source of organic compost used in farming and small scale sustainable, organic farming, the process of vermicasting is undergoing research as a treatment for organic waste in sewage and wastewater plants around the world.

Suitable Species

One of the earthworm species most often used for composting is the Red Wiggler (*Eisenia feti-da* or *Eisenia andrei*); *Lumbricus rubellus* (a.k.a. red earthworm or dilong (China)) is another breed of worm that can be used, but it does not adapt as well to the shallow compost bin as does *Eisenia fetida*. European nightcrawlers (*Eisenia hortensis*) may also be used. Users refer to European nightcrawlers by a variety of other names, including dendrobaenas, dendras, and Belgian nightcrawlers. African Nightcrawlers (*Eudrilus eugeniae*) are another set of popular composters. *Lumbricus terrestris* (a.k.a. Canadian nightcrawlers (US) or common earthworm (UK)) are not recommended, as they burrow deeper than most compost bins can accommodate.

Blueworms (*Perionyx excavatus*) may be used in the tropics.

These species commonly are found in organic-rich soils throughout Europe and North America and live in rotting vegetation, compost, and manure piles. They may be an invasive species in some areas. As they are shallow-dwelling and feed on decomposing plant matter in the soil, they adapt easily to living on food or plant waste in the confines of a worm bin.

Composting worms are available to order online, from nursery mail-order suppliers or angling shops where they are sold as bait. They can also be collected from compost and manure piles. These species are not the same worms that are found in ordinary soil or on pavement when the soil is flooded by water.

Large Scale

Large-scale vermicomposting is practiced in Canada, Italy, Japan, Malaysia, the Philippines, and the United States. The vermicompost may be used for farming, landscaping, to create compost tea, or for sale. Some of these operations produce worms for bait and/or home vermicomposting.

There are two main methods of large-scale vermiculture. Some systems use a windrow, which consists of bedding materials for the earthworms to live in and acts as a large bin; organic material is added to it. Although the windrow has no physical barriers to prevent worms from escaping, in theory they should not due to an abundance of organic matter for them to feed on. Often windrows are used on a concrete surface to prevent predators from gaining access to the worm population.

The windrow method and compost windrow turners were developed by Fletcher Sims Jr. of the Compost Corporation in Canyon, Texas. The Windrow Composting system is noted as a sustainable, cost-efficient way for farmers to manage dairy waste.

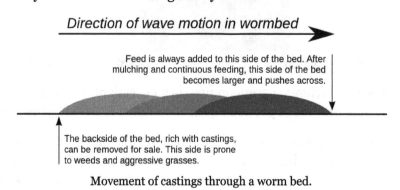

Movement of castings through a worm bed.

The second type of large-scale vermicomposting system is the raised bed or flow-through system. Here the worms are fed an inch of "worm chow" across the top of the bed, and an inch of castings are harvested from below by pulling a breaker bar across the large mesh screen which forms the base of the bed.

Because red worms are surface dwellers constantly moving towards the new food source, the flow-through system eliminates the need to separate worms from the castings before packaging. Flow-through systems are well suited to indoor facilities, making them the preferred choice for operations in colder climates.

Small Scale

For vermicomposting at home, a large variety of bins are commercially available, or a variety of adapted containers may be used. They may be made of old plastic containers, wood, Styrofoam, or metal containers. The design of a small bin usually depends on where an individual wishes to store the bin and how they wish to feed the worms.

Demonstration home scale worm bin at a community garden site - painted plywood

Some materials are less desirable than others in worm bin construction. Metal containers often conduct heat too readily, are prone to rusting, and may release heavy metals into the vermicompost. Styrofoam containers may release chemicals into the organic material. Some cedars, Yellow cedar, and Redwood contain resinous oils that may harm worms, although Western Red Cedar has excellent longevity in composting conditions. Hemlock is another inexpensive and fairly rot-resistant wood species that may be used to build worm bins.

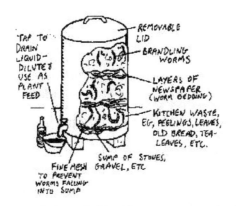

Diagram of a household-scale worm composting bin

Bins need holes or mesh for aeration. Some people add a spout or holes in the bottom for excess liquid to drain into a tray for collection. The most common materials used are plastic: recycled polyethylene and polypropylene and wood. Worm compost bins made from plastic are ideal, but require more drainage than wooden ones because they are non-absorbent. However, wooden bins will eventually decay and need to be replaced.

Small-scale vermicomposting is well-suited to turn kitchen waste into high-quality soil amendments, where space is limited. Worms can decompose organic matter without the additional human physical effort (turning the bin) that bin composting requires.

Composting worms which are detritivorous (eaters of trash), such as the red wiggler *Eisenia fetidae*, are epigeic (surface dwellers) together with symbiotic associated microbes are the ideal vectors for decomposing food waste. Common earthworms such as *Lumbricus terrestris* are anecic(deep burrowing) species and hence unsuitable for use in a closed system. Other soil species that contribute include insects, other worms and molds.

Climate and Temperature

There may be differences in vermicomposting methods depending on the climate. It is necessary to monitor the temperatures of large-scale bin systems (which can have high heat-retentive properties), as the feedstocks used can compost, heating up the worm bins as they decay and killing the worms.

The most common worms used in composting systems, redworms (*Eisenia foetida, Eisenia andrei,* and *Lumbricus rubellus*) feed most rapidly at temperatures of 15–25 °C (59-77 °F). They can survive at 10 °C (50 °F). Temperatures above 30 °C (86 °F) may harm them. This temperature range means that indoor vermicomposting with redworms is possible in all but tropical climates. Other worms like Perionyx excavatus are suitable for warmer climates. If a worm bin is kept outside, it should be placed in a sheltered position away from direct sunlight and insulated against frost in winter.

Feedstock

There are few food wastes that vermicomposting cannot compost, although meat waste and dairy products are likely to putrefy, and in outdoor bins can attract vermin. Green waste should be added in moderation to avoid heating the bin.

Small-Scale or Home Systems

Such systems usually use kitchen and garden waste, using "earthworms and other microorganisms to digest organic wastes, such as kitchen scraps". This includes:

- All fruits and vegetables (including citrus and other "high acid" foods)
- Vegetable and fruit peels and ends
- Coffee grounds and filters
- Tea bags (even those with high tannin levels)

- Grains such as bread, cracker and cereal (including moldy and stale)
- Eggshells (rinsed off)
- Leaves and grass clippings (not sprayed with pesticides)

Large-scale or Commercial

Such vermicomposting systems need reliable sources of large quantities of food. Systems presently operating use:

- Dairy cow or pig manure
- Sewage sludge
- Brewery waste
- Cotton mill waste
- Agricultural waste
- Food processing and grocery waste
- Cafeteria waste
- Grass clippings and wood chips

Harvesting

Vermicompost is ready for harvest when it contains few-to-no scraps of uneaten food or bedding. There are several methods of harvesting from small-scale systems: "dump and hand sort", "let the worms do the sorting", "alternate containers" and "divide and dump." These differ on the amount of time and labor involved and whether the vermicomposter wants to save as many worms as possible from being trapped in the harvested compost.

Worms in a bin being harvested

The pyramid method of harvesting worm compost is considered the simplest method for single layer bins. It is commonly used in small scale vermiculture.

While harvesting, it's also a good idea to try to pick out as many eggs/cocoons as possible and return them to the bin. Eggs are small, lemon-shaped yellowish objects that can usually be seen pretty easily with the naked eye and picked out.

Properties

Vermicompost has been shown to be richer in many nutrients than compost produced by other composting methods. It has also outperformed a commercial plant medium with nutrients added, but levels of magnesium required adjustment, as did pH.

However, in one study it has been found that homemade backyard vermicompost was lower in microbial biomass, soil microbial activity, and yield of a species of ryegrass than municipal compost,

It is rich in microbial life which converts nutrients already present in the soil into plant-available forms.

Unlike other compost, worm castings also contain worm mucus which helps prevent nutrients from washing away with the first watering and holds moisture better than plain soil.

Increases in the total nitrogen content in vermicompost, an increase in available nitrogen and phosphorus, as well as the increased removal of heavy metals from sludge and soil have been reported. The reduction in the bioavailability of heavy metals has been observed in a number of studies.

Benefits

Soil

- Improves soil aeration
- Enriches soil with micro-organisms (adding enzymes such as phosphatase and cellulase)
- Microbial activity in worm castings is 10 to 20 times higher than in the soil and organic matter that the worm ingests
- Attracts deep-burrowing earthworms already present in the soil
- Improves water holding capacity

Plant Growth

- Enhances germination, plant growth, and crop yield
- Improves root growth and structure
- Enriches soil with micro-organisms (adding plant hormones such as auxins and gibberellic acid)

Economic

- Biowastes conversion reduces waste flow to landfills

- Elimination of biowastes from the waste stream reduces contamination of other recyclables collected in a single bin (a common problem in communities practicing single-stream recycling)

- Creates low-skill jobs at local level

- Low capital investment and relatively simple technologies make vermicomposting practical for less-developed agricultural regions

Environmental

- Helps to close the "metabolic gap" through recycling waste on-site

- Large systems often use temperature control and mechanized harvesting, however other equipment is relatively simple and does not wear out quickly

- Production reduces greenhouse gas emissions such as methane and nitric oxide (produced in landfills or incinerators when not composted or through methane harvest)

As Fertilizer

Vermicompost can be mixed directly into the soil, or steeped in water and made into a worm tea by mixing some vermicompost in water, bubbling in oxygen with a small air pump, and steeping for a number of hours or days.

Mid-scale worm bin (1 m X 2.5 m up to 1 m deep), freshly refilled with bedding

The microbial activity of the compost is greater if it is aerated during this period. The resulting liquid is used as a fertilizer or sprayed on the plants.

The dark brown waste liquid, or leachate, that drains into the bottom of some vermicomposting systems as water-rich foods break down, is best applied back to the bin when added moisture is needed due to the possibility of phytotoxin content and organic acids that may be toxic to plants.

The pH, nutrient, and microbial content of these fertilizers varies upon the inputs fed to worms. Pulverized limestone, or calcium carbonate can be added to the system to raise the pH.

Troubleshooting

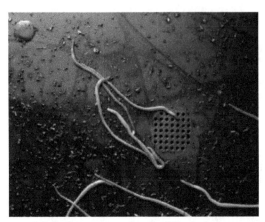

Worms and fruit fly pupas under the lid of a home worm bin.

Smells

When closed, a well-maintained bin is odorless; when opened, it should have little smell if any smell is present, it is earthy. Worms require gaseous oxygen. Oxygen can be provided by airholes in the bin, occasional stirring of bin contents, and removal of some bin contents if they become too deep or too wet. If decomposition becomes anaerobic from excess wet feedstock added to the bin, or the layers of food waste have become too deep, the bin will begin to smell of ammonia.

Moisture

If decomposition has become anaerobic, to restore healthy conditions and prevent the worms from dying, the smelly, excess waste water must be removed and the bin returned to a normal moisture level. To do this, first reduce addition of food scraps with a high moisture content and second, add fresh, dry bedding such as shredded newspaper to your bin, mixing it in well.

Pest Species

Pests such as rodents and flies are attracted by certain materials and odors, usually from large amounts of kitchen waste, particularly meat. Eliminating the use of meat or dairy product in a worm bin decreases the possibility of pests.

In warm weather, fruit and vinegar flies breed in the bins if fruit and vegetable waste is not thoroughly covered with bedding. This problem can be avoided by thoroughly covering the waste by at least 5 centimetres (2.0 in) of bedding. Maintaining the correct pH (close to neutral) and water content of the bin (just enough water where squeezed bedding drips a couple of drops) can help avoid these pests as well.

Worms Escaping

Worms generally stay in the bin, but may try to leave the bin when first introduced, or often after a rainstorm when outside humidity is high. Maintaining adequate conditions in the worm bin and putting a light over the bin when first introducing worms should eliminate this problem.

Nutrient Levels

Commercial vermicomposters test, and may amend their products to produce consistent quality and results. Because the small-scale and home systems use a varied mix of feedstocks, the nitrogen, potassium and phosphorus content of the resulting vermicompost will also be inconsistent. NPK testing may be helpful before the vermicompost or tea is applied to the garden.

In order to avoid over-fertilization issues, such as nitrogen burn, vermicompost can be diluted as a tea 50:50 with water, or as a solid can be mixed in 50:50 with potting soil.

The mucus produced creates a natural time-release fertilizer which cannot burn plants.

Windrow Composting

In agriculture, windrow composting is the production of compost by piling organic matter or biodegradable waste, such as animal manure and crop residues, in long rows (*windrows*). This method is suited to producing large volumes of compost. These rows are generally turned to improve porosity and oxygen content, mix in or remove moisture, and redistribute cooler and hotter portions of the pile. Windrow composting is a commonly used farm scale composting method. Composting process control parameters include the initial ratios of carbon and nitrogen rich materials, the amount of bulking agent added to assure air porosity, the pile size, moisture content, and turning frequency.

Windrow turner used on maturing piles at a biosolids composting facility in Canada.

Maturing windrows at an in-vessel composting facility.

The temperature of the windrows must be measured and logged constantly to determine the optimum time to turn them for quicker compost production.

Compost Windrow Turners

Compost windrow turners were developed to produce compost on a large scale by Fletcher Sims Jr. of Canyon, Texas. They are traditionally a large machine that straddles a windrow of 4 feet (1.25 meters) or more high, by as much as 12 feet (3.5 meters) across. Although smaller machines exist for small windrows, most operations use large machines for volume production. Turners drive through the windrow at a slow rate of forward movement. They have a steel drum with paddles that are rapidly turning. As the turner moves through the windrow, fresh air (oxygen) is injected into the compost by the drum/paddle assembly, and waste gases produced by bacterial decomposition are vented. The oxygen feeds the aerobic bacteria and thus speeds the composting process.

Utilization

To properly use a compost windrow turner, it is ideal to compost on a hard surfaced pad. Heavy-duty compost windrow turners allow the user to obtain optimum results with the aerobic hot composting process. By using four wheel drive or tracks the windrow turner is capable of turning compost in windrows located in remote locations. With a self-trailering option this allows the compost windrow turner to convert itself into a trailer to be pulled by a semi-truck tractor. These two options combined allow the compost windrow turner to be easily hauled anywhere and to work compost windrows in muddy and wet locations.

Specific Applications

Molasses-based distilleries all over the world generate large amount of effluent termed as spent wash or vinasse. For each liter of alcohol produced, around 8 liters of effluent is generated. This effluent has COD of 1,50,000 PPM and BOD of 60,000 PPM and even more. This effluent needs to be treated and the only effective method for conclusive disposal is by composting.

Sugar factories generate pressmud / cachaza during the process and the same has about 30% fibers as carbon and has large amounts of water. This pressmud is dumped on prepared land in the form of 100 meters long windrows of 3 meters x 1.5 meters and spent wash is sprayed on the windrow while the windrow is being turned. These machines help consume spent wash of about 2.5 times of the volume of the pressmud, which means that a 100 meters of windrow accommodates about 166 MT of pressmud and uses about 415 m³ of Spent wash in 50 days.

Microbial Culture (organic solution) TRIO COM-CULT is used about 1 kg per MT of pressmud for fast de-composing of the spent wash. Hundreds of thousands of square meters of spent wash is being composted all over the world in countries like India, Colombia, Brazil, Thailand, Indonesia, South Africa etc.

The compost yard has to be prepared in such a way that the land is impervious and does not allow the liquid effluent to pass down into the earth. The compost thus generated is of excellent quality and is rich in nutrients.

Garden Waste Dumping

Garden waste, or green waste dumping is the act of discarding or depositing garden waste somewhere it does not belong.

Garden waste is the accumulated plant matter from gardening activities which involve cutting or removing vegetation, i.e. cutting the lawn, weed removal, hedge trimming or pruning consisting of lawn clippings. leaf matter, wood and soil.

The composition and volume of garden waste can vary from season to season and location to location. A study in Aahrus, Denmark, found that on average, garden waste generation per person ranged between 122 kg to 155 kg per year.

Garden waste may be used to create compost or mulch, which can be used as a soil conditioner, adding valuable nutrients and building humus. The creation of compost requires a balance between, nitrogen, carbon, moisture and oxygen. Without the ideal balance, plant matter may take a long time to break down, drawing nitrogen from other sources, reducing nitrogen availability to existing vegetation which requires it for growth.

The risk of dumping garden waste is that it may contain seeds and plant parts that may grow (propagules), as well as increase fire fuel loads, disrupt visual amenity, accrue economic costs associated with the removal of waste as well as costs associated with the mitigation of associated impacts such as weed control, forest fire.

Cause

There are strong links between weed invasion of natural areas and the proximity and density of housing. The size and duration of the community have a direct relation to the density of weed infestation. Of the various means in which migration of exotic species from gardens take place, such as vegetative dispersal of runners, wind born and fallen seed, garden waste dumping can play a significant role. The results of one North German study found that of the problematic population of Fallopia, app. 29% originated from garden waste. Of a population of Heracleum mantegazzianum, 18% was found by Schepker to be generated by garden waste (as cited by Kowarik & von der Lippe, 2008) pg 24-25.

An Australia government publication suggest that some of the main reasons for the dumping of garden waste can be attributed to lack of care for the environment, convenience, or a reluctance to pay for the correct collection or disposal of the waste. (Environmental Protection Agency [EPA]. 2013). People dump garden waste to avoid disposal fees at landfill sites or because they do not want to spend the time or effort disposing of or recycling their waste properly. This activity is carried out by people in all parts of the community, from householders to businesses, such as professional landscapers and gardeners. The spread of exotic vegetation can out-compete locally endemic vegetation, altering the composition and structure of an ecosystem.

Dumping of garden waste in particular facilitates the spread of exotic vegetation into forest remnants via the introduction of seeds and propagules contained within the garden waste. Common selection criteria for home gardeners when choosing plants are often based on ease of propagation, suitability to local environmental conditions and novelty. These specific chosen characteristics increase the chance of plant parts and seeds that are introduced into forested areas becoming a problem.

The three major causes of animal habitat degradation are; the disruption or loss of ecosystem functions, the loss of food resources and loss of resident species. Non-native invaders can cause extinctions of vulnerable native species through competition, pest and disease transportation and habitat and ecosystem alteration.

The dumping of garden waste in nature reserves surrounding and near urban areas increases the risk of fires. The dumped garden waste will eventually dry out creating fuel adding to already fallen debris fuel load on which a fire can thrive and spread on. Garden waste can spread weeds and these weeds build fuel for fires. Dumped garden waste can facilitate higher rates of erosion by smothering natural vegetation cover. With no root systems for stabilisation the top soil is vulnerable to erosion (Ritter, J. 2015), This can add higher levels of sediments, contributing to the siltation of creeks and waterways.

If plant matter gets into waterways it can create reduced oxygen levels through the process of decomposition of green waste such as lawn clippings. This directly upsets the quality of water, affecting fish and aquatic wildlife. This dumping of green waste can also lead to the blocking of drainage systems; directly through the build-up of plant debris, and indirectly through the spread of invasive plant species that colonise wet areas, reducing and or changing the flow of waterways. This change in flow, including path and velocity, can alter hydrological cycles, affecting frequency and intensity of floods.

Impact

Increased Fire Risk

Dumping garden waste in nature reserves and parks surrounding and near urban areas can effect directly and indirectly the existing flora and fauna, as well as human life through the increased risk of fires. The dumped garden waste will eventually dry, creating additional fuel, adding to already fallen debris on which a fire can thrive and spread. Garden waste can spread weeds and these weeds also build fuel for fires. Fires may also spread to the suburban areas where humans can also be impacted by losing their homes from fire, incur injury or death from smoke or burns, and suffer economic loses such as income loss and clean-up costs. Fires can lead to an overall loss of habitat and biodiversity.

Threat to Biodiversity

The invasion of exotic plant species into remnant native forests is a threat to biodiversity. Some impacts of habitat degradation include; when native animals, insects and birds become vulnerable and put at risk; loss of food source for native wildlife; disruption of native plant-animal relationships ie pollination and seed dispersal and disconnection of plant-host relationships. Highly adaptive plants chosen for their ease of cultivation out compete more specialised species. Weed invasion of a forest system can change the processes of plant succession (the system of one species replacing another due to disturbance factors), the composition of the plant community and the composition and availability of nutrients. The change in forest composition can lead to loss of unique plant species. When a habitat is destroyed, the plants, animals, and other organisms that occupied the habitat have a reduced carrying capacity so that populations decline and extinction becomes a threat. Many endemic organisms have very specific requirements for their survival that can only be found within a certain ecosystem. The term 'hotspot' is used to describe areas featuring exceptional concentrations of endemic species and facing high potential of habitat degradation. The 25 most significant hotspots contain the habitats of 133,149 plant species (44% of all plant species worldwide; table 1) and 9,645 vertebrate species (35% of all vertebrate worldwide; table 2). These endemics are confined to an expanse of 2.1 million square kilometers (1.4% of land surface). Having lost 88% of their primary vegetation, they formerly occupied 17.4 million square kilometers or 11.8% of land surface. The recruitment of alien invasive species may lead to a homogenisation of landscapes. Although increased bio diversity in subregions created by newly introduced species may occur, the displacement of the existing plant species may lead to reduced biodiversity on a global scale.

Waterways Quality

This dumping of green waste such as lawn clippings can block drainage systems which pollutes water therefore effecting water quality and the health of aquatic plants and animals. Dumped garden waste can add high levels of sediments, reducing the light available for photosynthesis Dumping also block waterways and roads, cause flooding and facilitate higher rates of erosion by smothering natural vegetation cover.

Mitigation

Education on the value of biodiversity and the negative effects of weed invasion have been identified as key items in changing current trends. Specific education campaigns on the risks of dumping garden waste could be targeted at high-risk societal groups such as residents of housing in close proximity to reserves as well as members of gardening communities and plant sellers. Restricting the selection of garden species in new housing developments adjacent to reserves may reduce the effects of illegal dumping, thereby reducing requirement and associated cost of weed management. Creating habitat for wildlife by planting native plants, making a water source available, provide shelter and places to raise young. Healthy ecosystems are necessary for the survival and health of all organisms, and there are a number of ways to reduce negative impact on the environment. Cultivation of native plant species may benefit not only native plant populations but also native animal populations. For example, Sears & Anderson suggest that native bird species diversity in Australia and North America tend to match the volume and diversity of native vegetation. Crisp also explains the percentage of native insect species in a fauna has been found to be consistent with the percentage of native plant species.

Composting is a great way to recycle nutrients back into soils. Mulching the garden with leaves and clippings (BMCC, n.d). Fostering an appreciation of local natural environmental features and plant species may also help mitigate the issue. as well as the restriction of highly invasive plant species through international policy.

Utilization of green waste bins that are provided by some councils or shires that are emptied via curbside collection (BMCC. n.d). The addition of facilities for waste disposal could also improve the issue (DECC. 2008). Mitigation may involve governments holding campaigns that show people disposing legally and reporting the consequences for disposing illegally. A way Australian governments are addressing the problem is through the increase of fines in conjunction with better law enforcement. In Australia, fines can be up to $1,000,000 and can also incur imprisonment. The Protection of the Environment Operations Act imposes penalties for offences including polluting waters with waste, polluting land, illegally dumping waste or using land as an illegal waste facility.

Hydrothermal Liquefaction

Hydrothermal liquefaction (HTL) is a thermal depolymerization process used to convert wet biomass into crude-like oil -sometimes referred to as bio-oil or biocrude- under moderate temperature and high pressure. The crude-like oil (or bio-oil) has high energy density with a lower heating value of 33.8-36.9 MJ/kg and 5-20 wt% oxygen and renewable chemicals.

The reaction usually involves homogeneous and/or heterogeneous catalysts to improve the quality of products and yields. Carbon and hydrogen of an organic material, such as biomass, peat or low-ranked coals (lignite) are thermo-chemically converted into hydrophobic compounds with low viscosity and high solubility. Depending on the processing conditions, the fuel can be used as produced for heavy engines, including marine and rail or upgraded to transportation fuels, such as diesel, gasoline or jet-fuels.

History

As early as the 1920s, the concept of using hot water and alkali catalysts to produce oil out of biomass was proposed. This was the foundation of later HTL technologies that attracted research interest especially during the 1970s oil embargo. It was around that time that a high-pressure (hydrothermal) liquefaction process was developed at the Pittsburgh Energy Research Center (PERC) and later demonstrated (at the 100 kg/h scale) at the Albany Biomass Liquefaction Experimental Facility at Albany, Oregon, US. In 1982, Shell Oil developed the HTU™ process in the Netherlands. Other organizations that have previously demonstrated HTL of biomass include Hochschule für Angewandte Wissenschaften Hamburg, Germany, SCF Technologies in Copenhagen, Denmark, EPA's Water Engineering Research Laboratory, Cincinnati, Ohio, USA, and Changing World Technology Inc. (CWT), Philadelphia, Pennsylvania, USA. Today, technology companies such as Licella/Ignite Energy Resources (Australia), Altaca Energy (Turkey), Steeper Energy (Denmark, Canada), and Mulchand Holdings (India) continue to explore the commercialization of HTL.

Chemical Reactions

In hydrothermal liquefaction processes, long carbon chain molecules in biomass are thermally cracked and oxygen is removed in the form of H_2O (dehydration) and CO_2 (decarboxylation). These reactions result in the production of high H/C ratio bio-oil. Simplified descriptions of dehydration and decarboxylation reactions can be found in the literature (e.g. Asghari and Yoshida (2006) and Snåre et al. (2007))

Process

Most applications of hydrothermal liquefaction operate at temperatures between 250-550°C and high pressures of 5-25 MPa as well as catalysts for 20–60 minutes, although higher or lower temperatures can be used to optimize gas or liquid yields, respectively. At these temperatures and pressures, the water present in the biomass becomes either subcritical or supercritical, depending on the conditions, and acts as a solvent, reactant, and catalyst to facilitate the reaction of biomass to bio-oil.

The exact conversion of biomass to bio-oil is dependent on several variables:

- Feedstock composition
- Temperature and heating rate
- Pressure
- Solvent
- Residence time
- Catalysts

Feedstock

Theoretically, any biomass can be converted into bio-oil using hydrothermal liquefaction regardless of water content, and various different biomasses have been tested, from forestry and agriculture residues, sewage sludges, food process wasters, to emerging non-food biomass such as algae. The composition of cellulose, hemicellulose, protein, and lignin in the feedstock influence the yield and quality of the oil from the process.

Temperature and Heating Rate

Temperature plays a major role in the conversion of biomass to bio-oil. The temperature of the reaction determines the depolymerization of the biomass to bio-oil, as well as the repolymerization into char. While the ideal reaction temperature is dependent on the feedstock used, temperatures above ideal lead to an increase in char formation and eventually increased gas formation, while lower than ideal temperatures reduce depolymerization and overall product yields.

Similarly to temperature, the rate of heating plays a critical role in the production of the different phase streams, due to the prevalence of secondary reactions at non-optimum heating rates. Secondary reactions become dominant in heating rates that are too low, leading to the formation of char. While high heating rates are required to form liquid bio-oil, there is a threshold heating rate and temperature where liquid production is inhibited and gas production is favored in secondary reactions.

Pressure

Pressure (along with temperature) determines the super- or subcritical state of solvents as well as overall reaction kinetics and the energy inputs required to yield the desirable HTL products (oil, gas, chemicals, char etc.).

Residence Time

Hydrothermal liquefaction is a fast process, resulting in low residence times for depolymerization to occur. Typical residence times are measured in minutes (15 to 60 minutes); however, the residence time is highly dependent on the reaction conditions, including feedstock, solvent ratio and temperature. As such, optimization of the residence time is necessary to ensure a complete depolymerizaiton without allowing further reactions to occur.

Catalysts

While water acts as a catalyst in the reaction, other catalysts can be added to the reaction vessel to optimize the conversion. Previously used catalysts include water-soluble inorganic compounds and salts, including KOH and Na_2CO_3, as well as transition metal catalysts using Ni, Pd, Pt, and Ru supported on either carbon, silica or alumina. The addition of these catalysts can lead to an oil yield increase of 20% or greater, due to the catalysts converting the protein, cellulose, and hemicellulose into oil. This ability for catalysts to convert biomaterials other than fats and oils to bio-oil allows for a wider range of feedstock to be used .

Environmental Impact

Biofuels that are produced through hydrothermal liquefaction are carbon neutral, meaning that there are no net carbon emissions produced when burning the biofuel. The plant materials used to produce bio-oils use photosynthesis to grow, and as such consume carbon dioxide from the atmosphere. The burning of the biofuels produced releases carbon dioxide into the atmosphere, but is nearly completely offset by the carbon dioxide consumed from growing the plants, resulting in a release of only 15-18 g of CO_2/kWh or energy produced. This is substantially lower than the releases rate of fossil fuel technologies, which can range from releases of 955 g/kWh (coal), 813 g/kWh (oil), and 446 g/kWh (natural gas). Recently, Steeper Energy announced that the Carbon Intensity (CI) of its Hydrofaction™ oil is 15 CO_2eq/MJ according to GHGenius model (version 4.03a), while diesel fuel is 93.55 CO_2eq/MJ.

Hydrothermal liquefaction is a clean process that doesn't produce harmful compounds, such as ammonia, NO_x, or SO_x. Instead the heteroatoms, including nitrogen, sulfur, and chlorine, are converted into harmless byproducts such as N_2 and inorganic acids that can be neutralized with bases.

Compare with Pyrolysis and Other BTL Technologies

The HTL process differs from pyrolysis as it can process wet biomass and produce a bio-oil that contains approximately twice the energy density of pyrolysis oil. Pyrolysis is a related process to HTL, but biomass must be processed and dried in order to increase the yield. The presence of water in pyrolysis drastically increases the heat of vaporization of the organic material, increasing the energy required to decompose the biomass. Typical pyrolysis processes require a water content of less than 40% to suitably convert the biomass to bio-oil. This requires considerable pretreatment of wet biomass such as tropical grasses, which contain a water content as high as 80-85%, and even further treatment for aquatic species, which can contain higher than 90% water content.

The HTL oil can contain up to 80% of the feedstock carbon content (single pass). HTL oil has good potential to yield bio-oil with "drop-in" properties that can be directly distributed in existing petroleum infrastructure.

Mechanical Biological Treatment

A mechanical biological treatment (MBT) system is a type of waste processing facility that combines a sorting facility with a form of biological treatment such as composting or anaerobic digestion. MBT plants are designed to process mixed household waste as well as commercial and industrial wastes.

Process

The terms *mechanical biological treatment* or *mechanical biological pre-treatment* relate to a group of solid waste treatment systems. These systems enable the recovery of materials contained within the mixed waste and facilitate the stabilisation of the biodegradable component of the material.

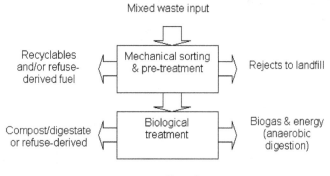

Process flow chart

The sorting component of the plants typically resemble a materials recovery facility. This component is either configured to recover the individual elements of the waste or produce a Refuse-derived fuel that can be used for the generation of power.

The components of the mixed waste stream that can be recovered include:

- Ferrous Metal
- Non-ferrous metal
- Plastic
- Glass

Terminology

MBT is also sometimes termed BMT – biological mechanical treatment – however this simply refers to the order of processing, i.e. the biological phase of the system precedes the mechanical sorting. MBT should not be confused with MHT – *mechanical heat treatment*

Mechanical Sorting

The "mechanical" element is usually an automated mechanical sorting stage. This either removes recyclable elements from a mixed waste stream (such as metals, plastics, glass and paper) or processes them. It typically involves factory style conveyors, industrial magnets, eddy current separators, trommels, shredders and other tailor made systems, or the sorting is done manually at hand picking stations. The mechanical element has a number of similarities to a materials recovery facility (MRF).

Wet material recovery facility, Hiriya, Israel

Some systems integrate a wet MRF to separate by density and floatation and to recover & wash the recyclable elements of the waste in a form that can be sent for recycling. MBT can alternatively process the waste to produce a high calorific fuel termed refuse derived fuel (RDF). RDF can be used in cement kilns or thermal combustion power plants and is generally made up from plastics and biodegradable organic waste. Systems which are configured to produce RDF include the Herhof and Ecodeco Processes. It is a common misconception that all MBT processes produce RDF. This is not the case and depends strictly on system configuration and suitable local markets for MBT outputs.

Biological Processing

The "biological" element refers to either:

- Anaerobic digestion
- Composting
- Biodrying

Anaerobic digestion harnesses anaerobic microorganisms to break down the biodegradable component of the waste to produce biogas and soil improver. The biogas can be used to generate electricity and heat.

Biological can also refer to a composting stage. Here the organic component is broken down by naturally occurring aerobic microorganisms. They breakdown the waste into carbon dioxide and compost. There is no green energy produced by systems employing only composting treatment for the biodegradable waste.

In the case of biodrying, the waste material undergoes a period of rapid heating through the action of aerobic microbes. During this partial composting stage the heat generated by the microbes result in rapid drying of the waste. These systems are often configured to produce a refuse-derived fuel where a dry, light material is advantageous for later transport and combustion.

Twin stage & UASB anaerobic digesters

Some systems incorporate both anaerobic digestion and composting. This may either take the form of a full anaerobic digestion phase, followed by the maturation (composting) of the digestate. Alternatively a partial anaerobic digestion phase can be induced on water that is percolated through the raw waste, dissolving the readily available sugars, with the remaining material being sent to a windrow composting facility.

By processing the biodegradable waste either by anaerobic digestion or by composting MBT technologies help to reduce the contribution of greenhouse gases to global warming.

Usable wastes for this system:

- Municipal solid waste

- Commercial and industrial waste

- Sewage sludge

Possible products of this system:

- Renewable fuel (biogas) leading to renewable power

- Recovered recyclable materials such as metals, paper, plastics, glass etc.

- Digestate - an organic fertiliser and soil improver

- Carbon credits – additional revenues

- High calorific fraction refuse derived fuel - Renewable fuel content dependent upon biological component

- Residual unusable materials prepared for their final safe treatment (e.g. incineration or gasification) and/or landfill

Further advantages:

- Small fraction of inert residual waste

- Reduction of the waste volume to be deposited to at least a half (density > 1.3 t/m^3), thus the lifetime of the landfill is at least twice as long as usually

- Utilisation of the leachate in the process

- Landfill gas not problematic as biological component of waste has been stabilised

- Daily covering of landfill not necessary

Consideration of Applications

MBT systems can form an integral part of a region's waste treatment infrastructure. These systems are typically integrated with kerbside collection schemes. In the event that a refuse-derived fuel is produced as a by-product then a combustion facility would be required. This could either be an incineration facility or a gasifier.

Alternatively MBT solutions can diminish the need for home separation and kerbside collection of recyclable elements of waste. This gives the ability of local authorities, municipalities and councils to reduce the use of waste vehicles on the roads and keep recycling rates high.

- Position of environmental groups

Friends of the Earth suggests that the best environmental route for residual waste is to firstly maximise removal of remaining recyclable materials from the waste stream (such as metals, plastics and paper). The amount of waste remaining should be composted or anaerobically digested and disposed of to landfill, unless sufficiently clean to be used as compost.

A report by Eunomia undertook a detailed analysis of the climate impacts of different residual waste technologies. It found that an MBT process that extracts both the metals and plastics prior to landfilling is one of the best options for dealing with our residual waste, and has a lower impact than either MBT processes producing RDF for incineration or incineration of waste without MBT.

Friends of the Earth does not support MBT plants that produce refuse derived fuel (RDF), and believes MBT processes should occur in small, localised treatment plants.

Trickling Filter

A trickling filter is a type of wastewater treatment system first used by Dibden and Clowes It consists of a fixed bed of rocks, lava, coke, gravel, slag, polyurethane foam, sphagnum peat moss, ceramic, or plastic media over which sewage or other wastewater flows downward and causes a layer of microbial slime (biofilm) to grow, covering the bed of media. Aerobic conditions are maintained by splashing, diffusion, and either by forced-air flowing through the bed or natural convection of air if the filter medium is porous.

The terms trickle filter, trickling biofilter, biofilter, biological filter and biological trickling filter are often used to refer to a trickling filter. These systems have also been described as roughing filters, intermittent filters, packed media bed filters, alternative septic systems, percolating filters, attached growth processes, and fixed film processes.

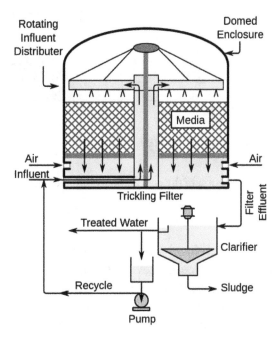

A typical complete trickling filter system

Construction

A typical trickling filter is circular and between 10 metres and 20 metres across and between 2 metres to 3 metres deep. A circular wall, often of brick, contains a bed of filter media which in turn rests on a base of under-drains. These under-drains function both to remove liquid passing through the filter media but also to allow the free passage of air up through the filter media. Mounted in the center over the top of the filter media is a spindle supporting two or more horizontal perforated pipes which extend to the edge of the media. The perforations on the pipes are designed to allow an even flow of liquid over the whole area of the media and are also angled so that when liquid flows from the pipes the whole assembly rotates around the central spindle. Settled sewage is delivered to a reservoir at the centre of the spindle via some form of dosing mechanism, often a tipping bucket device on small filters.

Larger filters may be rectangular and the distribution arms may be driven by hydraulic or electrical systems.

Trickling may have a variety of types of filter media used to support the bioi-film. Types of media most commonly used include coke, pumice, plastic matrix material, open-cell polyurethane foam, clinker, gravel, sand and geotextiles. Ideal filter medium optimizes surface area for microbial attachment, wastewater retention time, allows air flow, resists plugging is mechanically robust in all weathers allowing walking access across the filter and does not degrade. Some residential systems require forced aeration units which will increase maintenance and operational costs.

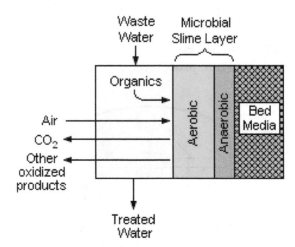

Image 1. A schematic cross-section of the contact face of the bed of media in a trickling filter

Operation

Typically, sewage flow enters at a high level and flows through the primary settlement tank. The supernatant from the tank flows into a dosing device, often a tipping bucket which delivers flow to the arms of the filter. The flush of water flows through the arms and exits through a series of holes pointing at an angle downwards. This propels the arms around distributing the liquid evenly over the surface of the filter media. Most are uncovered (unlike the accompanying diagram) and are freely ventilated to the atmosphere.

Systems can be configured for single-pass use where the treated water is applied to the trickling filter once before being disposed of, or for multi-pass use where a portion of the treated water is cycled back and re-treated via a closed loop. Multi-pass systems result in higher treatment quality and assist in removing Total Nitrogen (TN) levels by promoting nitrification in the aerobic media bed and denitrification in the anaerobic septic tank. Some systems use the filters in two banks operated in series so that the wastewater has two passes through a filter with a sedimentation stage between the two passes. Every few days the filters are switched round to balance the load. This method of treatment can improve nitrification and de-nitrification since much of the carbonaceous oxidative material is removed on the first pass through the filters.

The removal of pollutants from the waste water stream involves both absorption and adsorption of organic compounds and some inorganic species such as nitrite and nitrate ions by the layer of microbial bio film. The filter media is typically chosen to provide a very high surface area to volume. Typical materials are often porous and have considerable internal surface area in addition to the external surface of the medium. Passage of the waste water over the media provides dissolved oxygen which the bio-film layer requires for the biochemical oxidation of the organic compounds and releases carbon dioxide gas, water and other oxidized end products. As the bio film layer thickens, it eventually sloughs off into the liquid flow and subsequently forms part of the secondary sludge. Typically, a trickling filter is followed by a clarifier or sedimentation tank for the separation and removal of the sloughed film. Other filters utilizing higher-density media such as sand, foam and peat moss do not produce a sludge that must be removed, but require forced air blowers and backwashing or an enclosed anaerobic environment.

Biological Processes

The bio-film that develops in a trickling filter may become several millimetres thick and is typically a gelatinous matrix that contains many species of bacteria, cilliates and amoeboid protozoa, annelids, round worms and insect larvae and many other micro fauna. This is very different from many other bio-films which may be less than 1 mm thick. Within the thickness of the biofilm both aerobic and anaerobic zones can exist supporting both oxidative and reductive biological processes. At certain times of year, especially in the spring, rapid growth of organisms in the film may cause the film to be too thick and it may slough off in patches leading to the "spring slough".

Types

Single trickling filters may be used for the treatment of small residential septic tank discharges and very small rural sewage treatment systems. Larger centralized sewage treatment plants typically use many trickling filters in parallel. The treatment of industrial wastewater may involve specialised tricking filters which use plastic media and high flow rates.

Industrial Wastewater Treatment

Wastewaters from a variety of industrial processes have been treated in trickling filters. Such industrial wastewater trickling filters consist of two types:

- Large tanks or concrete enclosures filled with plastic packing or other media.

- Vertical towers filled with plastic packing or other media.

The availability of inexpensive plastic tower packings has led to their use as trickling filter beds in tall towers, some as high as 20 meters. As early as the 1960s, such towers were in use at: the Great Northern Oil's Pine Bend Refinery in Minnesota; the Cities Service Oil Company Trafalgar Refinery in Oakville, Ontario and at a kraft paper mill.

The treated water effluent from industrial wastewater trickling filters is typically processed in a clarifier to remove the sludge that sloughs off the microbial slime layer attached to the trickling filter media as for other trickling filter applications.

Some of the latest trickle filter technology involves aerated biofilters of plastic media in vessels using blowers to inject air at the bottom of the vessels, with either downflow or upflow of the wastewater.

Biofilter

Biofiltration is a pollution control technique using living material to capture and biologically degrade pollutants. Common uses include processing waste water, capturing harmful chemicals or silt from surface runoff, and microbiotic oxidation of contaminants in air.

Biosolids composting plant biofilter mound - note sprinkler visible front right to maintain proper moisture level for optimum functioning

Examples of biofiltration include;

- Bioswales, biostrips, biobags, bioscrubbers, and trickling filters
- Constructed wetlands and natural wetlands
- Slow sand filters
- Treatment ponds
- Green belts
- Green walls
- Riparian zones, riparian forests, bosques

Control of Air Pollution

When applied to air filtration and purification, biofilters use microorganisms to remove air pollution. The air flows through a packed bed and the pollutant transfers into a thin biofilm on the surface of the packing material. Microorganisms, including bacteria and fungi are immobilized in the biofilm and degrade the pollutant. Trickling filters and bioscrubbers rely on a biofilm and the bacterial action in their recirculating waters.

The technology finds greatest application in treating malodorous compounds and water-soluble volatile organic compounds (VOCs). Industries employing the technology include food and animal products, off-gas from wastewater treatment facilities, pharmaceuticals, wood products manufacturing, paint and coatings application and manufacturing and resin manufacturing and application, etc. Compounds treated are typically mixed VOCs and various sulfur compounds, including hydrogen sulfide. Very large airflows may be treated and although a large area (footprint) has typically been required—a large biofilter (>200,000 acfm) may occupy as much or more land than a football field—this has been one of the principal drawbacks of the technology. Engineered biofilters, designed and built since the early 1990s, have provided significant footprint reductions over the conventional flat-bed, organic media type.

One of the main challenges to optimum biofilter operation is maintaining proper moisture throughout the system. The air is normally humidified before it enters the bed with a watering (spray) system, humidification chamber, bioscrubber, or biotrickling filter. Properly maintained, a natural, organic packing media like peat, vegetable mulch, bark or wood chips may last for several years but engineered, combined natural organic and synthetic component packing materials will generally last much longer, up to 10 years. A number of companies offer these types or proprietary packing materials and multi-year guarantees, not usually provided with a conventional compost or wood chip bed biofilter.

Air cycle system at biosolids composting plant. Large duct in foreground is exhaust air into biofilter shown in next photo

Although widely employed, the scientific community is still unsure of the physical phenomena underpinning biofilter operation, and information about the microorganisms involved continues to be developed. A biofilter/bio-oxidation system is a fairly simple device to construct and operate and offers a cost-effective solution provided the pollutant is biodegradable within a moderate time frame (increasing residence time = increased size and capital costs), at reasonable concentrations (and lb/hr loading rates) and that the airstream is at an organism-viable temperature. For large volumes of air, a biofilter may be the only cost-effective solution. There is no secondary pollution (unlike the case of incineration where additional CO_2 and NO_x are produced from burning fuels) and degradation products form additional biomass, carbon dioxide and water. Media irrigation water, although many systems recycle part of it to reduce operating costs, has a moderately high biochemical oxygen demand (BOD) and may require treatment before disposal. However, this "blowdown water", necessary for proper maintenance of any bio-oxidation system, is generally accepted by municipal publicly owned treatment works without any pretreatment.

Biofilters are being utilized in Columbia Falls, Montana at Plum Creek Timber Company's fiberboard plant. The biofilters decrease the pollution emitted by the manufacturing process and the exhaust emitted is 98% clean. The newest, and largest, biofilter addition to Plum Creek cost $9.5 million, yet even though this new technology is expensive, in the long run it will cost less overtime than the alternative exhaust-cleaning incinerators fueled by natural gas (which are not as environmentally friendly). The biofilters use trillions of microscopic bacteria that cleanse the air being released from the plant.

Water Treatment

Biofiltration was first introduced in England in 1893 as a trickling filter for wastewater treatment and has since been successfully used for the treatment of different types of water. Biological treatment has been used in Europe to filter surface water for drinking purposes since the early 1900s and is now receiving more interest worldwide. Biofiltration is also common in wastewater treatment, aquaculture and greywater recycling as a way to minimize water replacement while increasing water quality.

Biofiltration Process

A biofilter is a bed of media on which microorganisms attach and grow to form a biological layer called biofilm. Biofiltration is thus usually referred to as a fixed–film process. Generally, the biofilm is formed by a community of different microorganisms (bacteria, fungi, yeast, etc.), macro-organisms (protozoa, worms, insect's larvae, etc.) and extracellular polymeric substances (EPS) (Flemming and Wingender, 2010). The aspect of the biofilm is usually slimy and muddy.

Biofilter installation at a commercial composting facility.

Water to be treated can be applied intermittently or continuously over the media, via upflow or downflow. Typically, a biofilter has two or three phases, depending on the feeding strategy (percolating or submerged biofilter):

- a solid phase (media);
- a liquid phase (water);
- a gaseous phase (air).

Organic matter and other water components diffuse into the biofilm where the treatment occurs, mostly by biodegradation. Biofiltration processes are usually aerobic, which means that microorganisms require oxygen for their metabolism. Oxygen can be supplied to the biofilm, either concurrently or countercurrently with water flow. Aeration occurs passively by the natural flow of air through the process (three phase biofilter) or by forced air supplied by blowers.

Microorganisms' activity is a key-factor of the process performance. The main influencing factors are the water composition, the biofilter hydraulic loading, the type of media, the feeding strategy (percolation or submerged media), the age of the biofilm, temperature, aeration, etc.

Types of Filtering Media

Originally, biofilter was developed using rock or slag as filter media, but different types of material are used today. These materials are categorized as inorganic media (sand, gravel, geotextile, different shapes of plastic media, glass beads, etc.) and organic media (peat, wood chips, coconut shell fragments, compost, etc.)

Advantages

Although biological filters have simple superficial structures, their internal hydrodynamics and the microorganisms' biology and ecology are complex and variable. These characteristics confer robustness to the process. In other words, the process has the capacity to maintain its performance or rapidly return to initial levels following a period of no flow, of intense use, toxic shocks, media backwash (high rate biofiltration processes), etc.

The structure of the biofilm protects microorganisms from difficult environmental conditions and retains the biomass inside the process, even when conditions are not optimal for its growth. Biofiltration processes offer the following advantages: (Rittmann et al., 1988):

- Because microorganisms are retained within the biofilm, biofiltration allows the development of microorganisms with relatively low specific growth rates;

- Biofilters are less subject to variable or intermittent loading and to hydraulic shock;

- Operational costs are usually lower than for activated sludge;

- Final treatment result is less influenced by biomass separation since the biomass concentration at the effluent is much lower than for suspended biomass processes;

- Attached biomass becomes more specialized (higher concentration of relevant organisms) at a given point in the process train because there is no biomass return.

Drawbacks

Because filtration and growth of biomass leads to an accumulation of matter in the filtering media, this type of fixed-film process is subject to clogging and flow channeling. Depending on the type of application and on the media used for microbial growth, clogging can be controlled using physical and/or chemical methods. Whenever possible, backwash steps can be implemented using air and/or water to disrupt the biomat and recover flow. Chemicals such as oxidizing (peroxide, ozone) or biocide agents can also be used.

Drinking Water

For drinking water, biological water treatment involves the use of naturally-occurring microorganisms in the surface water to improve water quality. Under optimum conditions, including relatively low turbidity and high oxygen content, the organisms break down material in the water and thus improve water quality. Slow sand filters or carbon filters are used to provide a support on which these microorganisms grow. These biological treatment systems effectively reduce water-borne diseases, dissolved organic carbon, turbidity and color in surface water, thus improving overall water quality.

Wastewater

Biofiltration is used to treat wastewater from a wide range of sources, with varying organic compositions and concentrations. Many examples of biofiltration applications are described in the literature. As a non-exhaustive list of applications, and notwithstanding the type of media, biofilters were developed and commercialized for the treatment of animal wastes, landfill leachates, dairy wastewater, domestic wastewater.

This process is versatile as it can be adapted to small flows (< 1 m3/d), such as onsite domestic wastewater as well as to flows generated by a municipality (> 240 000 m3/d). For decentralized domestic wastewater production, such as for isolated dwellings, it has been demonstrated that there are important daily, weekly and yearly fluctuations of hydraulic and organic production rates related to modern families' lifestyle. In this context, a biofilter located after a septic tank constitutes a robust process able to sustain the variability observed without compromising the treatment performance.

Use in Aquaculture

The use of biofilters is common in closed aquaculture systems, such as recirculating aquaculture systems (RAS). Many designs are used, with different benefits and drawbacks, however the function is the same: reducing water exchanges by converting ammonia to nitrate. Ammonia (NH_4^+ and NH_3) originates from the brachial excretion from the gills of aquatic animals and from the decomposition of organic matter. As ammonia-N is highly toxic, this is converted to a less toxic form of nitrite (by *Nitrosomonas* sp.) and then to an even less toxic form of nitrate (by *Nitrobacter* sp.). This "nitrification" process requires oxygen (aerobic conditions), without which the biofilter can crash. Furthermore, as this nitrification cycle produces H^+, the pH can decrease which necessitates the use of buffers such as lime.

References

- Diaz E (editor). (2008). Microbial Biodegradation: Genomics and Molecular Biology (1st ed.). Caister Academic Press. ISBN 1-904455-17-4. http://www.horizonpress.com/biod.

- Ratner, Buddy D. (2004). Pyrolytic carbon. In Biomaterials science: an introduction to materials in medicine. Academic Press. p. 171-180. ISBN 0-12-582463-7. Google Book Search. Retrieved 7 July 2011.

- C.P. Leslie Grady, Glenn T. Daigger and Henry C. Lim (1998). Biological wastewater Treatment (2nd ed.). CRC Press. ISBN 0-8247-8919-9.

- C.C. Lee & Shun Dar Lin (2000). Handbook of Environmental Engineering Calculations (1st ed.). McGraw Hill. ISBN 0-07-038183-6.

- Frank R. Spellman (2000). Spellman's Standard Handbook for Wastewater Operators. CRC Press. ISBN 1-56676-835-7.

- Appelhof, Mary (2007). Worms Eat My Garbage (2nd ed.). Kalamazoo, Mich.: Flowerfield Enterprises. ISBN 978-0-9778045-1-1.

- Tchobanoglous, G., Burton, F.L., and Stensel, H.D. (2003). Wastewater Engineering (Treatment Disposal Reuse) / Metcalf & Eddy, Inc. (4th ed.). McGraw-Hill Book Company. ISBN 0-07-041878-0.

- "Bioremediation of contaminated marine sediments can enhance metal mobility due to changes of bacterial diversity". Water Research. January 2015.

- Kenneth Mainord (September 1994). "Cleaning with Heat: Old Technology with a Bright New Future" (PDF). Pollution Prevention Regional Information Center. The Magazine of Critical Cleaning Technology. Retrieved 4 December 2015.

- Gary Davis & Keith Brown (April 1996). "Cleaning Metal Parts and Tooling" (PDF). Pollution Prevention Regional Information Center. Process Heating. Retrieved 4 December 2015.

- Thomas S. Dwan (2 September 1980). "Process for vacuum pyrolysis removal of polymers from various objects". Espacenet. European Patent Office. Retrieved 26 December 2015.

- "Paint Stripping: Reducing Waste and Hazardous Material". Minnesota Technical Assistance Program. University of Minnesota. July 2008. Retrieved 4 December 2015.

- Lago, Valentina, "Application of Mechanically Fluidized Reactors to Lignin Pyrolysis" (2015). Electronic Thesis and Dissertation Repository. Paper 2779.

- Ramirez, Jerome; Brown, Richard; Rainey, Thomas (1 July 2015). "A Review of Hydrothermal Liquefaction Bio-Crude Properties and Prospects for Upgrading to Transportation Fuels". Energies. 8 (7): 6765. doi:10.3390/en8076765.

- Sergios Karatzos; James D. McMillan; Jack N. Saddler (July 2014). "The Potential and Challenges of Drop-in Biofuels" (PDF). A report by IEA Bioenergy Task 39. Retrieved 3 Sep 2015.

- AARHUS UNIVERSITY (6 Feb 2013). "Hydrothermal liquefaction -- the most promising path to a sustainable bio-oil production". Retrieved 3 Sep 2015

- Olapade, OA; Ronk, AJ (2014). "Isolation, Characterization and Community Diversity of Indigenous Putative Toluene-Degrading Bacterial Populations with Catechol-2,3-Dioxygenase Genes in Contaminated Soils". Microbial Ecology. 69: 59–65. doi:10.1007/s00248-014-0466-6. PMID 25052383.

- Dongbing Li, Franco Berruti, Cedric Briens, "Autothermal fast pyrolysis of birch bark with partial oxidation in a fluidized bed reactor", Fuel, Volume 121, 1 April 2014, Pages 27-38

Uses of Biodegradable Material

The uses stated in this chapter are biodegradable athletic footwear, biodegradable bag, dry animal dung fuel and cook stove. The content strategically encompasses and incorporates the major uses of biodegradable minerals, providing a complete understanding.

Biodegradable Athletic Footwear

Biodegradable athletic footwear is footwear that uses biodegradable materials with the ability to compost at the end-of-life phase. Such materials include natural biodegradable polymers, synthetic biodegradable polymers, and biodegradable blends. The use of biodegradable materials is a long-term solution to landfill pollution that can significantly help protect the natural environment by replacing the synthetic, non-biodegradable polymers found in athletic footwear.

Problem of Non-Degradable Waste

The United States athletic shoe market is a $13 billion-per-year dollar industry that sells more than 350 million pairs of athletic shoes annually. The global footwear consumption has nearly doubled every twenty years, from 2.5 billion pairs in 1950 to more than 19 billion pairs of shoes in 2005. The increase in demand for athletic shoe products have progressively decreased the useful lives of shoes as a result of the rapid market changes and new consumer trends. A shorter life cycle of athletic footwear has begun to create non-degradable waste in landfills due to synthetic and other non-biodegradable materials used in production. The considerable growth in industrial production and consumption has made the athletic footwear industry face the environmental challenge of generated end-of-life waste.

Ethylene Vinyl Acetate Copolymer

The athletic shoe midsole is one of the main contributors that lead to a generation of end-of-life waste because it is composed of polymeric foams based on ethylene vinyl acetate (EVA). Ethylene vinyl acetate (EVA) is a polyolefin copolymer of ethylene and vinyl acetate that provides durability and flexibility, making it the most commonly used material found in athletic shoe midsoles. Although the synthetic polymer is a useful material for the athletic shoe industry, it has become an environmental concern because of its poor biodegradability. EVA goes through an anaerobic decomposition process called thermal degradation that often occurs in landfills resulting in releases of volatile organic compounds (VOC's) into the air. VOC's "contribute to the formation of tropospheric ozone, which is harmful to humans and plant life." Thermal degradation of EVA is temperature dependent and occurs in two stages; in the first stage acetic acid is lost, followed by the degradation of the unsaturated polyethylene polymer.

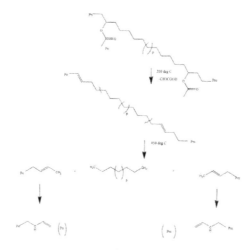

Thermal degradation of EVA by allylic scission.

Environmental Impact

The environmental impacts of athletic shoe degradation in landfills "are inextricably connected to the nature of the materials." The production of many petroleum-based products, such as EVA, used to manufacture athletic shoes result in serious environmental pollution of groundwater and rivers when disposed into landfills. When disposed in landfills, athletic footwear can take up to thousands of years to naturally degrade. EVA athletic shoe midsoles can be kept in contact with moist soil for a period of 12 years and experience little to no evidence of bio deterioration.

Although there are some that are taking initiatives to produce environmentally friendly athletic footwear, most of the footwear industry's response to this increasing problem of end-of-life shoe waste has been negligible. In order to reduce post-consumer waste and improve environmental properties of athletic shoes, biodegradable materials can help to replace synthetic polymers such as ethylene vinyl acetate with the ability to compost at the end-of-life phase.

Biodegradable Materials

"Biodegradation is a chemical degradation of materials provoked by the action of microorganisms such as bacteria, fungi, and algae." Although there are many materials categorized as biodegradable, there has been an increasing interest of biodegradable polymers that can lead to waste management options for polymers in the environment. These biodegradable polymers can be broken down into three categories: natural biodegradable polymer, synthetic biodegradable polymer, and biodegradable blends.

Natural Biodegradable Polymers

Natural biodegradable polymers are formed in nature during growth cycles of all organisms. When searching for natural fibers to replace synthetic materials in athletic shoes, the major natural biodegradable polymer that offers the most potential are polysaccharides. Starch is a polysaccharide that is useful because it readily degrades into harmless products when placed in contact with soil microorganisms.

Enzymatic hydrolysis of starch.

Starch is not often used alone as a plastic material because of its brittle nature, but is commonly used as a biodegradation additive. Many plasticizers use starch-glycerol-water to modify starch's brittle nature. Biodegradation of this blend was tested and was found that by the second day the degraded carbon had already attained about 100% of the initial carbon of the sample.

Synthetic Biodegradable Polymer

Hydrolytic Degradation of the aliphatic polyester, PLA.

Aliphatic polyesters are a diverse family of synthetic polymers of which are biocompatible, bio-degradable, and non-toxic. Specifically, poly (lactic acid) has low melt strength and low viscosity properties that are similar to EVA midsoles in athletic shoes. Poly (lactic acid) (PLA) is part of the poly (amide-enamine) group and can go through thermoplastic and foaming processes. Along with its good mechanical properties, its popularity is based on the non-toxic products that it becomes when it decomposes through hydrolytic degradation. Hydrolytic degradation of PLA generates the monomer lactic acid, which is metabolized via the tri-carboxylic acid cycle and eliminated as carbon dioxide.

Biodegradable Blends

Most synthetic polymers are resistant to microbial attack because of their physical and chemical properties. However, they can become biodegradable when introducing natural polymers such as starch. Natural polymers introduce ester groups that attach to the backbone of non-biodegradable polymers, making them more susceptible to degradation. Due to biodegradable polymers having limited properties; blending synthetic polymers can bring economic advantages and superior properties.

Insertion of an ester group into vinyl polymer.

End-of-Life Management

Although total elimination of post-consumer waste may not be feasible, proactive approaches to reduce the enormous amount of waste that 350 million pairs of athletic shoes create can make a difference in the environment. Biodegradable materials, such as biodegradable polymers, are a viable solution to aid in avoiding the end-of-life athletic footwear waste consumption. The major advantage of introducing biodegradable polymers to athletic footwear is the ability to compost with other organic wastes for it to become useful soil attendant products.

An alternative short-term approach to end-of-life management is recycling activities in the footwear industry. One major shoe manufacture, Nike Inc., created "Reuse-A-Shoe" program that involves recycling discarded athletic shoes by grinding and shredding the shoes to produce a material called "Nike Grind," which can be used in surfacing for tennis and basketball playgrounds or running tracks. Currently, the "Reuse-A-Shoe" program recycles approximately 125,000 pairs of shoes per year in the United States.

Recycling and composting are two major proposed solutions to end-of-life management. However, the use of biodegradable materials is a long-term solution that can significantly help protect the natural environment by replacing synthetic, non-biodegradable polymers found in athletic footwear.

Biodegradable Bag

Biodegradable bags are bags that are capable of being decomposed by bacteria or other living organisms.

The seal of a biodegradable bag in French

Every year approximately 500 billion to 1 trillion plastic bags are used worldwide.

Distinguishing "Biodegradable" from "Compostable"

In typical parlance, the word biodegradable is distinct in meaning from compostable. While biodegradable simply means an object is capable of being decomposed by bacteria or other living organisms, "compostable" in the plastic industry is defined as able to decompose in aerobic environments that are maintained under specific controlled temperature and humidity conditions. Compostable means capable of undergoing biological decomposition in a compost site such that the material is not visually distinguishable and breaks down into carbon dioxide, water, inorganic compounds and biomass at a rate consistent with known compostable materials. (ref: ASTM International D 6002)

The inclusion of "inorganic materials" precludes the end product from being considered as compost, or humus, which is purely organic material. Indeed, under the ASTM definition, the only criterion needed for a plastic to be called compostable is that it has to appear to go away at the same rate as something else that one already knows is compostable under the traditional definition.

Plastic bags can be made "oxo-biodegradable" by being manufactured from a normal plastic polymer (i.e. polyethylene) or polypropylene incorporating an additive which causes degradation and then biodegradation of the polymer (polyethylene) due to oxidation.

Trade Associations

The trade association for the Oxo-biodegradable plastics industry is the Oxo-biodegradable Plastics Association, which will certify products tested according to ASTM D6954 or (as from 1st Jan 2010) UAE 5009:2009

The trade associations for the compostable plastics industry are the Biodegradable Products Institute (BPI), "European Bioplastics," and SPIBioplastics Council. Plastics are certified as compostable for industrial composting conditions in the United States if they comply with ASTM D6400, and in Europe with the EN13432.

Companies

Different companies use different kinds of biodegradable bags. Many stores and companies are beginning to use different types of biodegradable bags to comply with environmental benefits. Multinational baking giant Grupo Bimbo SAB de CV of Mexico City claims to have been the first to make "Oxo Biodegradable metalized polypropylene snack bag". In addition to that, a company named "Doo Bandits" has created oxo- biodegradable bags used for picking up dog waste. The Supermarket Chain Aldi Süd in Germany offers bio-based Ecovio bags. Ecoflex bags are, like ordinary and oxo-bio bags flexible, tear-resistant, waterproof, and suitable for printing, but are much more expensive.

Materials

Most bags that are manufactured from plastic made from corn-based materials, like Polylactic acid (PLA). Biodegradable plastic bags require more plastic per bag, because the material is not as strong. Many bags are also made from paper, organic materials, or polycaprolactone.

"The public looks at biodegradable as something magical," even though the term is mostly meaningless, according to Ramani Narayan, a chemical engineer at Michigan State University in East Lansing, and science consultant to the Biodegradable Plastics Institute. "This is the most used and abused and misused word in our dictionary right now. Simply calling something biodegradable and not defining in what environment it is going to be biodegradable and in what time period it is going to degrade is very misleading and deceptive." In the Great Pacific garbage patch, biodegradable plastics break up into small pieces that can more easily enter the food chain by being consumed."

Recycling

In- plant scrap can often be recycled but post-consumer sorting and recycling is difficult. Bio-based polymers will contaminate the recycling of other more common polymers. While oxo-bio-degradable plastic manufacturers claim that their bags are recyclable, many plastic film recyclers will not accept them, as there have been no long-term studies on the viability of recycled-content products with these additives. Further, the Biodegradable Plastics Institute (BPI) says that the formulation of additives in oxo films varies greatly, which introduces even more variability in the recycling process. SPI Resin identification code 7 is applicable.

Marketing Qualification and Legal Issues

Since many of these plastics require access to sunlight, oxygen, or lengthy periods of time to achieve degradation or biodegradation, the Federal Trade Commission's GUIDES FOR THE USE OF ENVIRONMENTAL MARKETING CLAIMS, commonly called the "green guide," require proper marking of these products to show their performance limits.

The FTC provides an example:

Example 1: A trash bag is marketed as "degradable," with no qualification or other disclosure. The marketer relies on soil burial tests to show that the product will decompose in the presence of water and oxygen. The trash bags are customarily disposed of in incineration facilities or at sanitary landfills that are managed in a way that inhibits degradation by minimizing moisture and oxygen. Degradation will be irrelevant for those trash bags that are incinerated and, for those disposed of in landfills, the marketer does not possess adequate substantiation that the bags will degrade in a reasonably short period of time in a landfill. The claim is therefore deceptive

Since there are no pass-fail tests for "biodegradable" plastic bags, manufacturers must print on the product the environmental requirements for biodegradation to take place, time frame and end results in order to be within US Trade Requirements.

In 2007, the State of California essentially made the term "biodegradable bags" illegal, unless such terms are "substantiated by competent and reliable evidence to prevent deceiving or misleading consumers about environmental impact of degradable, compostable, and biodegradable plastic bags, food service ware, and packaging."

In 2010, an Australian manufacturer of plastic bags who made unsubstantiated or unqualified claims about biodegradability was fined by the Australian Competition and Consumer Commission (ACCC), which is the Australian equivalent of the FTC.

In recent years, the BPI (the certifying body for compostable plastic) and related companies have claimed products compost in available compost facilities at 60 °C (140 °F). The Vermont attorney general found these claims to be misleading and sued compostable plastic companies for false claims.

Dry Animal Dung Fuel

Dry animal dung fuel (or dry manure fuel) is animal feces that has been dried in order to be used as a fuel source. It is used as a fuel in many countries around the world. Using dry manure as a fuel source is an example of reuse of excreta. A disadvantage of using this kind of fuel is increased air pollution.

Stirling-Motor powered with cow dung in the Technical Collection Hochhut in Frankfurt on Main

Dry Dung and Moist Dung

Dry dung is more commonly used than moist dung, because it burns more easily. Dry manure is typically defined as having a moisture content less than 30 percent.

Benefits

The M.N. Yavari, of Peru built by Thames Iron Works, London in 1861-62 had a Watt steam engine (powered by dried lama dung) until 1914

The benefits of using dry animal dung include:

- Cheaper than most modern fuels
- Efficient
- Alleviates local pressure on wood resources
- Readily available - short walking time required to collect fuel
- No cash outlays necessary for purchase (can be exchanged for other products)
- Less environmental pollution
- Safer disposal of animal dung
- Sustainable and renewable energy source

Countries

Drying cow dung fuel

Africa

- In Egypt dry animal dung (from cows & buffaloes) is mixed with straw or crop residues to make dry fuel called "Gella" or "Jilla" dung cakes in modern times and ""khoroshtof"" in medieval times. Ancient Egyptians used the dry animal dung as a source of fuel. Dung cakes and building crop residues were the source of 76.4% of gross energy consumed in Egypt's rural areas during the 1980s. Temperatures of dung-fueled fires in an experiment on Egyptian village-made dung cake fuel produced

Egyptian women making "Gella" dry animal dung fuel

""a maximum of 640 degrees C in 12 minutes, falling to 240 degrees C after 25 minutes and 100 degrees C after 46 minutes. These temperatures were obtained without refueling and without bellows etc.""

Also, camel dung is used as fuel in Egypt.

- Lisu is the cakes of dry cow dung fuel in Lesotho (see photo)

Huts in a village near Maseru, Lesotho. The fuel being used on the fire is dried cattle dung

- Mali

Asia

Dung cooking fire. Pushkar India.

- China

Water buffalo dung fuel drying on a wall in a Hani ethnic minority village in Yuanyang county, Yunnan, China

- Nepal

- Iran since prehistoric time to modern eras

- In India dry buffalo dung is used as fuel and it is sometimes a sacred practice to use cow dung fuel in some areas in India. Cow dung is known as """Gomaya""" or """Komaya""" in India. Dry animal dung cakes are called Upla in Hindi.

- In Pakistan cow/buffalo dung is used as fuel.

- Bangladesh dry cow dung fuel is called Ghunte.

- Afghanistan

- Kyrgyz Republic Dung is used in specially designed home stoves, which vent to the outside

U.S. soldiers patrolling outside a qalat covered in caked and dried cow dung in an Afghani village

Cow dung fuel was burnt on the Gauchar's Historical Field, India to gauge the direction of air currents

Making Komaya (cow dung fuel in India)

Europe

- Russians dry animal dung is known as ""Kiziak"" which is made by collecting dried animal dung on the steppe, wetting it in water then mixing it with straw then making it in discs which were then dried in the sun. It was used as a source of fuel for the winter and, throughout the summer.

- France in Maison du Marais poitevin in Coulon there is a demonstration of traditional usage of dry dung fuel.

Dung cakes being prepared for fuel on the Ile de Brehat, Brittany, France, c. 1900.

The Americas

- Early European settlers on the Great Plains of the United States used dried buffalo manure as a fuel. They called it buffalo chips.

- American officials in Texas are studying using dry cow dung as a fuel

- Pueblo Indians used dry animal dung as a fuel

- In Peru, the Yavari steam ship was fueled by Lama dung fuel for several decades.

- Dry dung can be used in the production of celluloid for film.

Human Feces

Human feces can in principal also be dried and used as a fuel source if they are collected in a type of dry toilet, for example an incinerating toilet. Since 2011, the Bill & Melinda Gates Foundation is supporting the development of such toilets as part of their "Reinvent the Toilet Challenge" to promote safer, more effective ways to treat human excreta. The omni-processor is another example of using human feces contained in faecal sludge or sewage sludge as a fuel source.

History

Dry animal dung was used from prehistoric times, including in Ancient Persia and Ancient Egypt. In Equatorial Guinea archaeological evidence has been found of the practice and biblical records indicate animal and human dung were used as fuel.

Biogas

Biogas typically refers to a mixture of different gases produced by the breakdown of organic matter in the absence of oxygen. Biogas can be produced from raw materials such as agricultural waste, manure, municipal waste, plant material, sewage, green waste or food waste. Biogas is a renewable energy source and in many cases exerts a very small carbon footprint.

Biogas can be produced by anaerobic digestion with anaerobic organisms, which digest material inside a closed system, or fermentation of biodegradable materials.

Biogas can be produced by anaerobic digestion with anaerobic organisms, which digest material inside a closed system, or fermentation of biodegradable materials.

Pipes carrying biogas (foreground), natural gas and condensate

Biogas is primarily methane ($CH4$) and carbon dioxide (CO_2) and may have small amounts of hydrogen sulfide ($H2S$), moisture and siloxanes. The gases methane, hydrogen, and carbon monoxide (CO) can be combusted or oxidized with oxygen. This energy release allows biogas to be used as a fuel; it can be used for any heating purpose, such as cooking. It can also be used in a gas engine to convert the energy in the gas into electricity and heat.

Biogas can be compressed, the same way natural gas is compressed to CNG, and used to power motor vehicles. In the UK, for example, biogas is estimated to have the potential to replace around 17% of vehicle fuel. It qualifies for renewable energy subsidies in some parts of the world. Biogas can be cleaned and upgraded to natural gas standards, when it becomes bio-methane. Biogas is considered to be a renewable resource because its production-and-use cycle is continuous, and it generates no net carbon dioxide. Organic material grows, is converted and used and then regrows in a continually repeating cycle. From a carbon perspective, as much carbon dioxide is absorbed from the atmosphere in the growth of the primary bio-resource as is released when the material is ultimately converted to energy.

Production

Biogas production in rural Germany

Biogas is produced as landfill gas (LFG), which is produced by the breakdown of Biodegradable waste inside a landfill due to chemical reactions and microbes, or as digested gas, produced inside an anaerobic digester. A *biogas plant* is the name often given to an anaerobic digester that treats farm wastes or energy crops. It can be produced using anaerobic digesters (air-tight tanks with different configurations). These plants can be fed with energy crops such as maize silage or biodegradable wastes including sewage sludge and food waste. During the process, the microorganisms transform biomass waste into biogas (mainly methane and carbon dioxide) and digestate. The biogas is a renewable energy that can be used for heating, electricity, and many other operations that use a reciprocating internal combustion engine, such as GE Jenbacher or Caterpillar gas engines. Other internal combustion engines such as gas turbines are suitable for the conversion of biogas into both electricity and heat. The digestate is the remaining inorganic matter that was not transformed into biogas. It can be used as an agricultural fertiliser.

There are two key processes: mesophilic and thermophilic digestion which is dependent on temperature. In experimental work at University of Alaska Fairbanks, a 1000-litre digester using psychrophiles harvested from "mud from a frozen lake in Alaska" has produced 200–300 liters of methane per day, about 20%–30% of the output from digesters in warmer climates.

Dangers

The dangers of biogas are mostly similar to those of natural gas, but with an additional risk from the toxicity of its hydrogen sulfide fraction. Biogas can be explosive when mixed in the ratio of one part biogas to 8-20 parts air. Special safety precautions have to be taken for entering an empty biogas digester for maintenance work.

It is important that a biogas system never has negative pressure as this could cause an explosion. Negative gas pressure can occur if too much gas is removed or leaked; Because of this biogas should not be used at pressures below one column inch of water, measured by a pressure gauge.

Frequent smell checks must be performed on a biogas system. If biogas is smelled anywhere windows and doors should be opened immediately. If there is a fire the gas should be shut off at the gate valve of the biogas system.

Landfill Gas

Landfill gas is produced by wet organic waste decomposing under anaerobic conditions in a biogas.

The waste is covered and mechanically compressed by the weight of the material that is deposited above. This material prevents oxygen exposure thus allowing anaerobic microbes to thrive. This gas builds up and is slowly released into the atmosphere if the site has not been engineered to capture the gas. Landfill gas released in an uncontrolled way can be hazardous since it can become explosive when it escapes from the landfill and mixes with oxygen. The lower explosive limit is 5% methane and the upper is 15% methane.

The methane in biogas is 20 times more potent a greenhouse gas than carbon dioxide. Therefore, uncontained landfill gas, which escapes into the atmosphere may significantly contribute to the effects of global warming. In addition, volatile organic compounds (VOCs) in landfill gas contribute to the formation of photochemical smog.

Technical

Biochemical oxygen demand (BOD) is a measure of the amount of oxygen required by aerobic micro-organisms to decompose the organic matter in a sample of water. Knowing the energy density of the material being used in the biodigester as well as the BOD for the liquid discharge allows for the calculation of the daily energy output from a biodigester.

Another term related to biodigesters is effluent dirtiness, which tells how much organic material there is per unit of biogas source. Typical units for this measure are in mg BOD/litre. As an example, effluent dirtiness can range between 800–1200 mg BOD/litre in Panama.

From 1 kg of decommissioned kitchen bio-waste, 0.45 m³ of biogas can be obtained. The price for collecting biological waste from households is approximately €70 per ton.

Composition

Typical composition of biogas		
Compound	Formula	%
Methane	CH_4	50–75
Carbon dioxide	CO_2	25–50
Nitrogen	N_2	0–10
Hydrogen	H_2	0–1
Hydrogen sulfide	H_2S	0–3
Oxygen	O_2	0–0.5
Source: *www.kolumbus.fi, 2007*		

The composition of biogas varies depending upon the origin of the anaerobic digestion process. Landfill gas typically has methane concentrations around 50%. Advanced waste treatment technologies can produce biogas with 55%–75% methane, which for reactors with free liquids can be increased to 80%-90% methane using in-situ gas purification techniques. As produced, biogas contains water vapor. The fractional volume of water vapor is a function of biogas temperature; correction of measured gas volume for water vapor content and thermal expansion is easily done via simple mathematics which yields the standardized volume of dry biogas.

In some cases, biogas contains siloxanes. They are formed from the anaerobic decomposition of materials commonly found in soaps and detergents. During combustion of biogas containing siloxanes, silicon is released and can combine with free oxygen or other elements in the combustion gas. Deposits are formed containing mostly silica (SiO_2) or silicates (Si_xO_y) and can contain calcium, sulfur, zinc, phosphorus. Such *white mineral* deposits accumulate to a surface thickness of several millimeters and must be removed by chemical or mechanical means.

Practical and cost-effective technologies to remove siloxanes and other biogas contaminants are available.

For 1000 kg (wet weight) of input to a typical biodigester, total solids may be 30% of the wet weight while volatile suspended solids may be 90% of the total solids. Protein would be 20% of the volatile solids, carbohydrates would be 70% of the volatile solids, and finally fats would be 10% of the volatile solids.

Benefits of Manure Derived Biogas

High levels of methane are produced when manure is stored under anaerobic conditions. During storage and when manure has been applied to the land, nitrous oxide is also produced as a byproduct of the denitrification process. Nitrous oxide (N2O) is 320 times more aggressive as a greenhouse gas than carbon dioxide and methane 25 times more than carbon dioxide.

By converting cow manure into methane biogas via anaerobic digestion, the millions of cattle in the United States would be able to produce 100 billion kilowatt hours of electricity, enough to power millions of homes across the United States. In fact, one cow can produce enough manure in one day to generate 3 kilowatt hours of electricity; only 2.4 kilowatt hours of electricity are needed to power a single 100-watt light bulb for one day. Furthermore, by converting cattle manure into methane biogas instead of letting it decompose, global warming gases could be reduced by 99 million metric tons or 4%.

Applications

Biogas can be used for electricity production on sewage works, in a CHP gas engine, where the waste heat from the engine is conveniently used for heating the digester; cooking; space heating; water heating; and process heating. If compressed, it can replace compressed natural gas for use in vehicles, where it can fuel an internal combustion engine or fuel cells and is a much more effective displacer of carbon dioxide than the normal use in on-site CHP plants.

A biogas bus in Linköping, Sweden

Biogas Upgrading

Raw biogas produced from digestion is roughly 60% methane and 29% CO2 with trace elements of H2S; it is not of high enough quality to be used as fuel gas for machinery. The corrosive nature of H2S alone is enough to destroy the internals of a plant.

Methane in biogas can be concentrated via a biogas upgrader to the same standards as fossil natural gas, which itself has to go through a cleaning process, and becomes *biomethane*. If the local gas network allows, the producer of the biogas may use their distribution networks. Gas must be very clean to reach pipeline quality and must be of the correct composition for the distribution network to accept. Carbon dioxide, water, hydrogen sulfide, and particulates must be removed if present.

There are four main methods of upgrading: water washing, pressure swing adsorption, selexol adsorption, and amine gas treating. In addition to these, the use of membrane separation technology for biogas upgrading is increasing, and there are already several plants operating in Europe and USA.

The most prevalent method is water washing where high pressure gas flows into a column where the carbon dioxide and other trace elements are scrubbed by cascading water running counter-flow to the gas. This arrangement could deliver 98% methane with manufacturers guaranteeing maximum 2% methane loss in the system. It takes roughly between 3% and 6% of the total energy output in gas to run a biogas upgrading system.

Biogas Gas-Grid Injection

Gas-grid injection is the injection of biogas into the methane grid (natural gas grid). Injections includes biogas until the breakthrough of micro combined heat and power two-thirds of all the energy produced by biogas power plants was lost (the heat), using the grid to transport the gas to customers, the electricity and the heat can be used for on-site generation resulting in a reduction of losses in the transportation of energy. Typical energy losses in natural gas transmission systems range from 1% to 2%. The current energy losses on a large electrical system range from 5% to 8%.

Biogas in Transport

"Biogaståget Amanda" ("The Biogas Train Amanda") train near Linköping station, Sweden

If concentrated and compressed, it can be used in vehicle transportation. Compressed biogas is becoming widely used in Sweden, Switzerland, and Germany. A biogas-powered train, named Biogaståget Amanda (The Biogas Train Amanda), has been in service in Sweden since 2005. Biogas powers automobiles. In 1974, a British documentary film titled *Sweet as a Nut* detailed the biogas production process from pig manure and showed how it fueled a custom-adapted combustion engine. In 2007, an estimated 12,000 vehicles were being fueled with upgraded biogas worldwide, mostly in Europe.

Measuring in Biogas Environments

Biogas is part of the wet gas and condensing gas (or air) category that includes mist or fog in the gas stream. The mist or fog is predominately water vapor that condenses on the sides of pipes or stacks throughout the gas flow. Biogas environments include wastewater digesters, landfills, and animal feeding operations (covered livestock lagoons).

Ultrasonic flow meters are one of the few devices capable of measuring in a biogas atmosphere. Most thermal flow meters are unable to provide reliable data because the moisture causes steady high flow readings and continuous flow spiking, although there are single-point insertion thermal mass flow meters capable of accurately monitoring biogas flows with minimal pressure drop. They can handle moisture variations that occur in the flow stream because of daily and seasonal temperature fluctuations, and account for the moisture in the flow stream to produce a dry gas value.

Legislation

European Union

The European Union has legislation regarding waste management and landfill sites called the Landfill Directive.

Countries such as the United Kingdom and Germany now have legislation in force that provides farmers with long-term revenue and energy security.

United States

The United States legislates against landfill gas as it contains VOCs. The United States Clean Air Act and Title 40 of the Code of Federal Regulations (CFR) requires landfill owners to estimate the quantity of non-methane organic compounds (NMOCs) emitted. If the estimated NMOC emissions exceeds 50 tonnes per year, the landfill owner is required to collect the gas and treat it to remove the entrained NMOCs. Treatment of the landfill gas is usually by combustion. Because of the remoteness of landfill sites, it is sometimes not economically feasible to produce electricity from the gas.

Global Developments

United States

With the many benefits of biogas, it is starting to become a popular source of energy and is starting to be used in the United States more. In 2003, the United States consumed 147 trillion BTU of energy from "landfill gas", about 0.6% of the total U.S. natural gas consumption. Methane biogas derived from cow manure is being tested in the U.S. According to a 2008 study, collected by the *Science and Children* magazine, methane biogas from cow manure would be sufficient to produce 100 billion kilowatt hours enough to power millions of homes across America. Furthermore, methane biogas has been tested to prove that it can reduce 99 million metric tons of greenhouse gas emissions or about 4% of the greenhouse gases produced by the United States.

In Vermont, for example, biogas generated on dairy farms was included in the CVPS Cow Power program. The program was originally offered by Central Vermont Public Service Corporation as a voluntary tariff and now with a recent merger with Green Mountain Power is now the GMP Cow Power Program. Customers can elect to pay a premium on their electric bill, and that premium is passed directly to the farms in the program. In Sheldon, Vermont, Green Mountain Dairy has provided renewable energy as part of the Cow Power program. It started when the brothers who own the farm, Bill and Brian Rowell, wanted to address some of the manure management challenges faced by dairy farms, including manure odor, and nutrient availability for the crops they need to grow to feed the animals. They installed an anaerobic digester to process the cow and milking center waste from their 950 cows to produce renewable energy, a bedding to replace sawdust, and a plant-friendly fertilizer. The energy and environmental attributes are sold to the GMP Cow Power program. On average, the system run by the Rowells produces enough electricity to power 300 to 350 other homes. The generator capacity is about 300 kilowatts.

In Hereford, Texas, cow manure is being used to power an ethanol power plant. By switching to methane biogas, the ethanol power plant has saved 1000 barrels of oil a day. Over all, the power plant has reduced transportation costs and will be opening many more jobs for future power plants that will rely on biogas.

In Oakley, Kansas, an ethanol plant considered to be one of the largest biogas facilities in North America is using Integrated Manure Utilization System "IMUS" to produce heat for its boilers by utilizing feedlot manure, municipal organics and ethanol plant waste. At full capacity the plant is expected to replace 90% of the fossil fuel used in the manufacturing process of ethanol.

Europe

The level of development varies greatly in Europe. While countries such as Germany, Austria and Sweden are fairly advanced in their use of biogas, there is a vast potential for this renewable energy source in the rest of the continent, especially in Eastern Europe. Different legal frameworks, education schemes and the availability of technology are among the prime reasons behind this untapped potential. Another challenge for the further progression of biogas has been negative public perception.

Initiated by the events of the gas crisis in Europe during December 2008, it was decided to launch the EU project "SEBE" (Sustainable and Innovative European Biogas Environment) which is financed under the CENTRAL programme. The goal is to address the energy dependence of Europe by establishing an online platform to combine knowledge and launch pilot projects aimed at raising awareness among the public and developing new biogas technologies.

In February 2009, the European Biogas Association (EBA) was founded in Brussels as a non-profit organisation to promote the deployment of sustainable biogas production and use in Europe. EBA's strategy defines three priorities: establish biogas as an important part of Europe's energy mix, promote source separation of household waste to increase the gas potential, and support the production of biomethane as vehicle fuel. In July 2013, it had 60 members from 24 countries across Europe.

UK

As of September 2013, there are about 130 non-sewage biogas plants in the UK. Most are on-farm, and some larger facilities exist off-farm, which are taking food and consumer wastes.

On 5 October 2010, biogas was injected into the UK gas grid for the first time. Sewage from over 30,000 Oxfordshire homes is sent to Didcot sewage treatment works, where it is treated in an anaerobic digestor to produce biogas, which is then cleaned to provide gas for approximately 200 homes.

In 2015 the Green-Energy company Ecotricity announced their plans to build three grid-injecting digester's.

Germany

Germany is Europe's biggest biogas producer and the market leader in biogas technology. In 2010 there were 5,905 biogas plants operating throughout the country: Lower Saxony, Bavaria, and the eastern federal states are the main regions. Most of these plants are employed as power plants. Usually the biogas plants are directly connected with a CHP which produces electric power by burning the bio methane. The electrical power is then fed into the public power grid. In 2010, the total installed electrical capacity of these power plants was 2,291 MW. The electricity supply was approximately 12.8 TWh, which is 12.6% of the total generated renewable electricity.

Biogas in Germany is primarily extracted by the co-fermentation of energy crops (called 'NawaRo', an abbreviation of *nachwachsende Rohstoffe*, German for renewable resources) mixed with manure. The main crop used is corn. Organic waste and industrial and agricultural residues such as waste from the food industry are also used for biogas generation.In this respect, biogas production in Germany differs significantly from the UK, where biogas generated from landfill sites is most common.

Biogas production in Germany has developed rapidly over the last 20 years. The main reason is the legally created frameworks. Government support of renewable energy started in 1991 with the Electricity Feed-in Act (*StrEG*). This law guaranteed the producers of energy from renewable sources the feed into the public power grid, thus the power companies were forced to take all produced energy from independent private producers of green energy. In 2000 the Electricity Feed-in Act was replaced by the Renewable Energy Sources Act (*EEG*). This law even guaranteed a fixed compensation for the produced electric power over 20 years. The amount of around 8 ¢/kWh gave farmers the opportunity to become energy suppliers and gain a further source of income.

The German agricultural biogas production was given a further push in 2004 by implementing the so-called NawaRo-Bonus. This is a special payment given for the use of renewable resources, that is, energy crops. In 2007 the German government stressed its intention to invest further effort and support in improving the renewable energy supply to provide an answer on growing climate challenges and increasing oil prices by the 'Integrated Climate and Energy Programme'.

This continual trend of renewable energy promotion induces a number of challenges facing the management and organisation of renewable energy supply that has also several impacts on the biogas production. The first challenge to be noticed is the high area-consuming of the biogas electric power supply. In 2011 energy crops for biogas production consumed an area of circa 800,000 ha in Germany. This high demand of agricultural areas generates new competitions with the food industries that did not exist hitherto. Moreover, new industries and markets were created in predominately rural regions entailing different new players with an economic, political and civil background. Their influence and acting has to be governed to gain all advantages this new source of energy is offering. Finally biogas will furthermore play an important role in the German renewable energy supply if good governance is focused.

Subcontinent

Biogas in India has been traditionally based on dairy manure as feed stock and these "gobar" gas plants have been in operation for a long period of time, especially in rural India. In the last 2-3 decades, research organisations with a focus on rural energy security have enhanced the design of the systems resulting in newer efficient low cost designs such as the Deenabandhu model.

The Deenabandhu Model is a new biogas-production model popular in India. (*Deenabandhu* means "friend of the helpless.") The unit usually has a capacity of 2 to 3 cubic metres. It is constructed using bricks or by a ferrocement mixture. In India, the brick model costs slightly more than the ferrocement model; however, India's Ministry of New and Renewable Energy offers some subsidy per model constructed.

LPG (Liquefied Petroleum Gas) is a key source of cooking fuel in urban India and its prices have been increasing along with the global fuel prices. Also the heavy subsidies provided by the successive governments in promoting LPG as a domestic cooking fuel has become a financial burden renewing the focus on biogas as a cooking fuel alternative in urban establishments. This has led to the development of prefabricated digester for modular deployments as compared to RCC and cement structures which take a longer duration to construct. Renewed focus on process technology like the Biourja process model has enhanced the stature of medium and large scale anaerobic digester in India as a potential alternative to LPG as primary cooking fuel.

In India, Nepal, Pakistan and Bangladesh biogas produced from the anaerobic digestion of manure in small-scale digestion facilities is called gobar gas; it is estimated that such facilities exist in over 2 million households in India, 50,000 in Bangladesh and thousands in Pakistan, particularly North Punjab, due to the thriving population of livestock. The digester is an airtight circular pit made of concrete with a pipe connection. The manure is directed to the pit, usually straight from the cattle shed. The pit is filled with a required quantity of wastewater. The gas pipe is connected to the kitchen fireplace through control valves. The combustion of this biogas has very little odour or smoke. Owing to simplicity in implementation and use of cheap raw materials in villages, it is one of the most environmentally sound energy sources for rural needs. One type of these system is the Sintex Digester. Some designs use vermiculture to further enhance the slurry produced by the biogas plant for use as compost.

To create awareness and associate the people interested in biogas, the Indian Biogas Association was formed. It aspires to be a unique blend of nationwide operators, manufacturers and planners of biogas plants, and representatives from science and research. The association was founded in 2010 and is now ready to start mushrooming. Its motto is "propagating Biogas in a sustainable way".

In Pakistan, the Rural Support Programmes Network is running the Pakistan Domestic Biogas Programme which has installed 5,360 biogas plants and has trained in excess of 200 masons on the technology and aims to develop the Biogas Sector in Pakistan.

In Nepal, the government provides subsidies to build biogas plant.

China

The Chinese had experimented the applications of biogas since 1958. Around 1970, China had installed 6,000,000 digesters in an effort to make agriculture more efficient. During the last years the technology has met high growth rates. This seems to be the earliest developments in generating biogas from agricultural waste.

In Developing Nations

Domestic biogas plants convert livestock manure and night soil into biogas and slurry, the fermented manure. This technology is feasible for small-holders with livestock producing 50 kg manure per day, an equivalent of about 6 pigs or 3 cows. This manure has to be collectable to mix it with water and feed it into the plant. Toilets can be connected. Another precondition is the temperature that affects the fermentation process. With an optimum at 36 C° the technology especially applies for those living in a (sub) tropical climate. This makes the technology for small holders in developing countries often suitable.

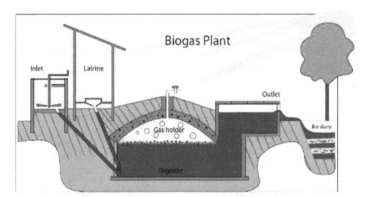

Simple sketch of household biogas plant

Depending on size and location, a typical brick made fixed dome biogas plant can be installed at the yard of a rural household with the investment between US$300 to $500 in Asian countries and up to $1400 in the African context. A high quality biogas plant needs minimum maintenance costs and can produce gas for at least 15–20 years without major problems and re-investments. For the user, biogas provides clean cooking energy, reduces indoor air pollution, and reduces the time needed for traditional biomass collection, especially for women and children. The slurry is a clean organic fertilizer that potentially increases agricultural productivity.

Domestic biogas technology is a proven and established technology in many parts of the world, especially Asia. Several countries in this region have embarked on large-scale programmes on domestic biogas, such as China and India.

The Netherlands Development Organisation, SNV, supports national programmes on domestic biogas that aim to establish commercial-viable domestic biogas

sectors in which local companies market, install and service biogas plants for households. In Asia, SNV is working in Nepal, Vietnam, Bangladesh, Bhutan, Cambodia, Lao PDR, Pakistan and Indonesia, and in Africa; Rwanda, Senegal, Burkina Faso, Ethiopia, Tanzania, Uganda, Kenya, Benin and Cameroon.

In South Africa a prebuilt Biogas system is manufactured and sold. One key feature is that installation requires less skill and is quicker to install as the digester tank is premade plastic.

Society and Culture

In the 1985 Australian film *Mad Max Beyond Thunderdome* the post-apocalyptic settlement Barter town is powered by a central biogas system based upon a piggery. As well as providing electricity, methane is used to power Barter's vehicles.

"Cow Town", written in the early 1940s, discuss the travails of a city vastly built on cow manure and the hardships brought upon by the resulting methane biogas. Carter McCormick, an engineer from a town outside the city, is sent in to figure out a way to utilize this gas to help power, rather than suffocate, the city.

Cook Stove

A cook stove is heated by burning wood, charcoal, animal dung or crop residue. Cook stoves are commonly used for cooking and heating food in developing countries.

Developing countries consume significantly less energy than developed country; however, over 50% of the energy is for cooking food. The average rural family spends 20% or more of its income purchasing wood or charcoal for cooking. The urban poor also frequently spend a significant portion of their income on the purchase of wood or charcoal.

Stove manufacture in Senegal.

Cooking over an open fire can cause increased health problems brought on from the smoke, particularly lung and eye ailments, but also birth defects. Replacing the traditional 3-rock cook stove with an improved one and venting the smoke out of the house through a chimney can significantly improve a family's health.

Built-in wood stove cooking food

Deforestation and erosion often the result from harvesting wood for cooking fuel. The main goal of most improved cooking stoves is to reduce the pressure placed on local forests by reducing the amount of wood the stoves consume.

Three-Stone Cooking Fire

The traditional method of cooking is on a three-stone cooking fire. It is the cheapest stove to produce, requiring only three suitable stones of the same height on which a cooking pot can be balanced over a fire. However, this cooking method also has problems:

cooking tortillas (flatbread) on a wood-fired 3-stone stove

- Smoke is vented into the home, instead of outdoors, causing health problems. According to the World Health Organization, "4.3 million people a year die prematurely from illness attributable to the household air pollution caused by the inefficient use of solid fuels (2012 data)."

- Fuel is wasted, as heat is allowed to escape into the open air. This requires the user to gather more fuel and may result in increased deforestation if wood is used for fuel.

- Only one cooking pot can be used at a time.

- The use of an open fire creates a risk of burns and scalds. Particularly when the stove is used indoors, cramped conditions make adults and particularly children susceptible to falling or stepping into the fire and receiving burns. Additionally, accidental spills of boiling water may result in scalding, and blowing on the fire to supply oxygen may discharge burning embers and cause eye injuries.

Improved Stoves and Other Measures

The World Health Organization has documented the significant number of deaths caused by smoke from home fires. The negative impacts can be reduced by using improved cook stoves, improved fuels (e.g. biogas, or kerosene instead of dung), changes to the environment (e.g. use of a chimney), and changes to user behavior (e.g. drying fuel wood before use, using a lid during cooking)." Improved stoves are more efficient, meaning that the stove's users spend less time gathering wood or other fuels, suffer less emphysema and other lung diseases prevalent in smoke-filled homes, while reducing deforestation and air pollution. However, a closed stove may result in production of more soot and ultra-fine particles than an open fire would.

For instance, an improvement of the energy efficiency from 25% for the traditional Lao stove (wood and charcoal fired) to 29% efficiency for an improved Lao stove, results in 21% less wood fuel being needed, and saves 182,000 ton CO2eq emissions, as reported in a GHG-compensation project. The traditional Lao stove needs an average of 385 kilograms of charcoal and 450 kilograms of wood per year as fuel. For making a kilogram of charcoal about seven kilograms of wood is needed. The report also indicates that in Cambodia 369,000 tons of non-renewable biomass wood fuel is consumed yearly for charcoal production for these stoves, destroying 45 km² of deciduous forests. 4% of the forest regrows.

Some designs also make the stove safer, preventing burns that often occur when children stumble into open fires. Some of the new stove designs are as follows:

Brick and Mortar Stove

A variety of new brick and mortar stoves have emerged. Most of the new designs incorporate a combustion chamber found in a Rocket stove. By confining combustion to an insulated and enclosed area, the stoves increase the core temperature and thereby achieve more complete combustion. This allows a smaller amount of fuel to burn hotter, while producing less ash and smoke. The Justa Stove is a simple biomass stove built around an insulated, elbow-shaped combustion chamber which provides more intense heat and cleaner combustion than an open fire, meaning that it consumes less fuel than a three-stone stove.

An improved Justa Stove jointly developed by the non-profit Proyecto Mirador and the Aprovecho Research Center called the "Dos por Tres" is being disseminated in Honduras with more than 20,000 stoves installed as of 2011. The Dos por Tres has been registered as Project 690 and certified by the Gold Standard Foundation to reduce greenhouse gas emissions by 2.7 tons per year. All

Proyecto Mirador documentation related to proof of these reductions can be viewed on the Gold Standard Project registry.

Dos por Tres Stove

The Justa Stove has been deployed in Honduras by Trees, Water & People and ADESHA, for which they jointly won an Ashden Award in 2005. The Eco Stove and the Patsari stove share common benefits with the Justa Stove, and are also used in Central America. Their proponents claim that these stoves use approximately 1/3 of the fuel required by traditional three stone stoves, lessening the daily labor devoted to gathering wood and also preventing deforestation. At the same time, it employs a stove pipe flue to vent fumes through the roof. This almost eliminates cooking smoke within the home, preventing respiratory problems for the users. Various groups have run programs to provide such stoves, or encourage production of stove making facilities, including certain Rotary Clubs; Trees, Water and People; and organizations aimed at preserving wildlife by preventing deforestation.

Lorena Adobe Stove

A predecessor to the Justa/Eco/Patsari Stoves was the Lorena adobe Stove. It was designed as a simple-to-build cook stove for use in Central America, that could be manufactured locally of materials. The name of Lorena Stove comes from the combination of the two Spanish words *lodo* and *arena* (meaning mud and sand) as the stoves were a combination of the two. The Lorena Stove is an enclosed stove of rammed earth construction, with a chimney built onto it.

The Lorena Stove was designed with the mistaken belief that rammed earth would act as insulation; there was a basic misunderstanding of the difference between mass and insulation. Good insulation resists the passage of heat; thermal mass does the opposite, it absorbs heat. Testing has shown that the rammed earth used in the Lorena stove absorbs heat that should be directed toward cooking.

However, this "waste" heat radiates into the structure, providing more efficient heating than an open fire (a disadvantage in hot weather). A Lorena stove can also be used to dry clothes as its mass slowly cools off after the fire dies.

The designers, Aprovecho, now state: "The Lorena has been tested over the years by many researchers and has generally been found to use more firewood than an indoor open fire. The stove has other attributes. Its chimney takes smoke out of the kitchen and it is well liked. It is pretty and a nice addition to the house. It is low cost and can be repaired and even built by the home owner. But, it is not a fuel saving or low emission stove". In later designs, the rammed earth has been replaced with thermal insulation, such as pumice or ash.

Kenya Ceramic Jiko

From the beginning of the appropriate technology movement, one of the principal goals has been to create an affordable stove that was more efficient than the universally used three stone cooking fire. The Kenya Ceramic Jiko is one of the improved stoves.

Charcoal is the standard cooking fuel in East Africa. Traditionally it was burned in a metal stove or "Jiko" as stoves are called in the Swahili language. The KCJ is the traditional Jiko with a ceramic liner. The initial model has a distinctive shape, differing from the traditional cylindrical jiko, with the top and bottom the same diameter, tapering at about 30 degrees to a waist.

Sanjha Chulha/Earth Stove/Surya Stove

Since 1999, an engineering company named Nishant Bioenergy (P) Limited in North India is conceiving, designing, fabricating and selling patent pending biomass briquette and pellet fueled cook stoves. These are useful for commercial cooking and are designed to burn biomass pellet and briquette. Biomass briquettes can be made from any farm or forest residues with or without binders and pellets are made with or without the addition of binder.

The company has won national and international recognition for their efforts, including the Ashden Award in 2005, PCRA Award-2001, and UN promising practices-2006.

Nishant Bioenergy developed its first stove in 1999 and named Sanjha Chulha (community-cooking stove) is an "Institutional Cook Stove" for use with multiple cooking pot and hot water tank as combined cooking. The stove name is derived from a village tradition where women would cook all their chappatis/rotis (Indian bread) on a *Sanjha Chulha*, which was a communal earthen cook stove used by everyone. The stove was designed by Ramesh K. Nibhoria who also designed the Earth Stove for cooking and frying use with a single pot. Earth Stove has two model, one is fixed pellet feed and second one with automated pellet feeding system. Earth Stove with automated pellet feeding system is very innovative and has controls on heat intensity from zero to full like LPG. New design is applied for patents and also being registered under copy write acts.

The Surya Stove by Nishant Bioenergy runs on powdered biomass such as saw dust or any other biomass that has a moisture content level of 15% . The stove has an automated fuel feed system (duly linked with temperature controller) and ash cleansing system. It has a temperature controller to pre-set the temperature of the cooking/frying medium. It is especially designed for the frying needs of the ready-to-eat food industry as well as for controlled hot air requirement.

Project Surya

Project Surya, by Scripps Institute of Oceanography at UCSD, field tested improved cook stoves and modifications during its pilot phase which is now complete and the results have been posted online. Project Surya has also launched the Carbon Credit Pilot Project (C2P2) to explore if rewarding women directly with funds from carbon markets, for using improved stoves, will significantly enhance adoption of the field tested stoves.

Prefab Stoves

The Ecocina stove was designed by StoveTeam International and is manufactured at a central location from cement, pumice, and ceramic tiles. It resembles a large flower pot, with a steel cooking surface which can also receive a pot. It was created by a volunteer worker who noticed the high number of respiratory illnesses and burns on patients in Guatemala. It is actively produced in countries such as Guatemala, El Salvador, Honduras, Nicaragua and Mexico. Unlike its brick and mortar counterparts, the Ecocina stoves have no flue and are manufactured in a backyard factory. They are then placed in a home on top of a table or similar raised surface. As with its brick and mortar counterpart, the Ecocina stove employs a rocket stove combustion chamber and promises reduction in the consumption of firewood and in the amount of fumes emitted into the home. It also remains cool to the touch, preventing burns.

Baker Stoves

The Baker cook stove was developed by Top Third Ventures and designed by Claesson Koivisto Rune. It is designed to emphasize aesthetic appeal, usability, and cultural conformity. Rural households in Kenya are its target consumer group. The Baker cook stove is made up of metal components. Thermal insulation and a forced air flow mechanism result in a higher combustion temperature and safer and cleaner cooking compared to the traditional three-stone cooking fire.

Turbo Stoves

Some new metal stoves employ turbo-charging features such that air pressed into the stove or swirled, will then significantly increase the efficiency of combustion.

The Lucia Stove developed by World Stove employs swirling air patterns to change combustion and has been economically produced. It is marketed by World Stove as one part of a larger environmental solution because it captures carbon and thereby reduces the amount of carbon in the atmosphere from cooking. It produces Biocarb that is then recycled back to the soil.

The Turbococina Stove was developed in El Salvador by René M. N. Suarez; the name is derived from the term "Turbocombustión" which is a new combustion method in which, during combustion, temperatures as low as possible are maintained to inhibit the formation of pollutants like NOx; an alternative is to reduce the concentration of oxygen below the stoichiometric requirement. The Turbococina promised positive results by employing higher pressures to lower combustion temperatures, but its high cost of production (stainless steel) and its use of electricity have prevented it from going into production. At present, it does not appear to be economically viable.

BioLite Stoves

The BioLite HomeStove was made to replace open cooking fires. Its design converts the heat into usable electricity to power a fan, which reduces fuel needs by 50%, toxic smoke by about 95%, and black carbon emissions by 91%. The amount of CO_2 saved per year by one stove is equivalent to the amount saved by buying a hybrid car. The effects of deforestation are lessened and time is regained by people spending less hours gathering wood for open fires. The remaining off-grid energy that does not power the fan can then be used to charge portable devices through a USB port, such as cell phones and LED lights.

The BioLite CampStove

Solar Stoves

Simple solar stoves use the following basic principles:

- Concentrating sunlight: A reflective mirror of polished glass, metal or metallised film concentrates light and heat from the sun on a small cooking area, making the energy more concentrated and increasing its heating power.

- Converting light to heat: A black or low reflectivity surface on a food container or the inside of a solar cooker improves the effectiveness of turning light into heat. Light absorption converts the sun's visible light into heat, substantially improving the effectiveness of the cooker.

- Trapping heat: Convection can be reduced by isolating the air inside the cooker from the air outside the cooker. A plastic bag or tightly sealed glass cover traps the hot air inside. This makes it possible to reach temperatures on cold and windy days similar to those possible on hot days.

Solar stove used for cooking dishes

Bio Ethanol Fueled Cook Stoves

Clean energy bio-ethanol cooking fuel in Kenya has been pioneered by International Research & Development Africa Ltd with a BIOMOTO Cook stove. Bio-ethanol fuel is manufactured from food crop stover and post harvest, contaminated and damaged starch crops purchased from bottom of pyramid farmers.

Another initiative is Project Gaia. Alcohol fuels, such as ethanol, burn quickly, cleanly and are renewable. Project Gaia works with the CleanCook, a clean-burning stove that use absorption technology to burn the ethanol.

Improved Cook Stoves

Improved Cook Stoves (ICS) are designed to reduce the fuel consumption per meal and to curb smoke emissions from open fires inside dwellings. They are designed for developing country settings as a low cost bridging technology.

There are various designs, such as the ONIL Stove which uses mortar-less concrete blocks in its construction and costs $150 USD per stove. Another design is the Berkeley-Darfur stove that reduces smoke and is twice as efficient as a clay stove, with the goal of reducing the need for women to leave the camps in search of wood.

The Save80 is a portable stove with an integral 8-liter pot (both made of stainless steel) and weighs about 4 kg. It has a nominal effective thermal power of 1.5 kW and needs 250 g of small dry wood sticks to bring 6 liters of water to the boil, which is 80% less than a traditional cooking fire. This efficiency is achieved using the 8-liter pot that is design-optimized for this stove. On one side near the upper rim there is a small port for feeding additional fuel into the already burning stove. The stove design ensures pre-heating of the air supply from the bottom inlet and complete combustion with no visible smoke and relatively small amounts of ash. Exhaust air outlets are on the side opposite the fuel feed port.

The Save80-Wonderbox is made of expanded polypropylene, which is specifically designed to contain the 8-liter pot of the Save80 stove. It is designed to keep a near boiling temperature for a relatively long time. After reaching the boiling temperature, food – for instance rice – can be transferred to the Wonderbox, a retained heat cooking device. There it continues to heat up.

The solar cooker uses no fuels. These devices require clear sunlight, and are practical in many of the sunny regions of the world.

The Energy and Resources Institute cookstove model SPT-610 has an efficiency of 37% and has been developed in collaboration with Indian Institute of Technology Delhi.

Advanced Biomass Cookstoves

There are two primary types of advanced biomass stoves that can achieve high levels of performance; forced air stoves and gasifier stoves, both of which can run on processed or raw biomass.

- Forced air stoves have a fan powered either by a battery, an external source of electricity, or a thermoelectric generator. This fan blows high velocity, low volume jets of air into the combustion chamber, which when optimized results in more complete combustion of the fuel.

- Gasifier stoves force the gases and smoke that result from incomplete combustion of fuels such as biomass back into the cookstove's flame, where the heat of the flame then continues to combust the particles in the smoke until almost complete combustion has occurred, reducing emissions. Typical gasifier stoves are known as Top Lit Updraft (TLUD) stoves because some fuel is lit on top of the stove, forcing combustible products to pass through the flame front before being emitted into the air.

Classification of Cook Stoves

Classification of cook stoves

1. Three-stone fire
2. Early "ICS" "Improved Cook Stoves" to 1990s (clay/ceramic/buckets)
3. Fuel-controlled stoves (mainly Rocket stoves)
   a. Simple (portable) b. Stationary (w/ chimney) c. Forced-air (FA)
4. Semi-gasifiers (mainly China and Vesto) w/ some air control
5. Gasifiers ("micro-" for cooking), some with FA (Fan Assistance)
   a. Top-lit updraft (known as TLUDs) w/ migrating pyrolytic zone (batch)
   b. Updrafts and downdrafts w/ stationary gasification zones (continuous)
   c. Other drafts, including cross and opposite/opposing drafts
6. "Fan-jet" with very strong air currents into the fuel (3 known examples):
   a. Philips-FA
   b. Lucia-FA
   c. Turbococina
7. Non-biomass. Stoves not using raw dry biomass fuels:
   Charcoal; alcohol; refined fossils; coal; biogas; electric; solar.

Multi-mode capable stoves can be used only in one way at any one time.

Refuse-Derived Fuel

Refuse-derived fuel (RDF) or solid recovered fuel / specified recovered fuel (SRF) is a fuel produced by shredding and dehydrating solid waste (MSW) with a Waste converter technology. RDF consists largely of combustible components of municipal waste such as plastics and biodegradable waste. RDF processing facilities are normally located near a source of MSW and, while an optional combustion facility is normally close to the processing facility, it may also be located at a remote location. SRF can be distinguished from RDF in the fact that it is produced to reach a standard such as CEN/343 ANAS. A comprehensive review is now available on SRF / RDF production, quality standards and thermal recovery, including statistics on European SRF quality.

Processing

Non-combustible materials such as glass and metals are removed during the post-treatment processing cycle with an air knife or other density separation technique. The residual material can be sold in its processed form (depending on the process treatment) or it may be compressed into Pellet fuel, bricks or logs and used for other purposes either stand-alone or in a recursive recycling process.

RDF is extracted from municipal solid waste using a mix of mechanical and/or biological treatment methods.

The production of RDF may involve the following steps:

- Bag splitting/Shredding
- Size screening
- Magnetic separation
- Coarse shredding
- Refining separation

The mix of materials should be cut to particles not bigger than 25 mm / 1 inch. The moisture content of the mix should not be bigger than 15% because if so the materials will not burn well in the furnace. The particles will last between 17 and 18 seconds in the furnace usually in a temperature of 1,200 °C / 2,192 °F.

End Markets

RDF can be used in a variety of ways to produce electricity. It can be used alongside traditional sources of fuel in coal power plants. In Europe RDF can be used in the cement kiln industry, where the strict standards of the Waste Incineration Directive are met. RDF can also be fed into plasma arc gasification modules, pyrolysis plants and where the RDF is capable of being combusted cleanly or in compliance with the Kyoto Protocol, RDF can provide a funding source where unused carbon credits are sold on the open market via a carbon exchange. However, the use of municipal waste contracts and the bankability of these solutions is still a relatively new concept, thus RDF's financial advantage may be debatable.

Measurement of the Biomass Fraction of RDF and SRF

The biomass fraction of RDF and SRF has a monetary value under multiple greenhouse gas protocols, such as the European Union Emissions Trading Scheme and the Renewable Obligation Certificate program in the United Kingdom. Biomass is considered to be carbon-neutral since the CO_2

liberated from the combustion of biomass is recycled in plants. The combusted biomass fraction of RDF/SRF is used by stationary combustion operators to reduce their overall reported CO_2 emissions.

Several methods have been developed by the European CEN 343 working group to determine the biomass fraction of RDF/SRF. The initial two methods developed (CEN/TS 15440) were the manual sorting method and the selective dissolution method. Since each method suffered from limitations in properly characterizing the biomass fraction, an alternative method was developed using the principles of radiocarbon dating. A technical review (CEN/TR 15591:2007) outlining the carbon-14 method was published in 2007, and a technical standard of the carbon dating method (CEN/TS 15747:2008) was published in 2008. In the United States, there is already an equivalent carbon-14 method under the standard method ASTM D6866.

Although carbon-14 dating can determine with excellent precision the biomass fraction of RDF/SRF, it cannot determine directly the biomass calorific value. Determining the calorific value is important for green certificate programs such as the Renewable Obligation Certificate program in the United Kingdom. These programs award certificates based on the energy produced from biomass. Several research papers, including the one commissioned by the Renewable Energy Association in the UK, have been published that demonstrate how the carbon-14 result can be used to calculate the biomass calorific value.

Regional Use

Iowa

The first full-scale waste-to-energy facility in the US was the Arnold O. Chantland Resource Recovery Plant, built in 1975 located in Ames, Iowa. This plant also produces RDF that is sent to a local power plant for supplemental fuel.

Manchester

The city of Manchester, in the north west of England, is in the process of awarding a contract for the use of RDF which will be produced by proposed mechanical biological treatment facilities as part of a huge PFI contract. The Greater Manchester Waste Disposal Authority has recently announced there is significant market interest in initial bids for the use of RDF which is projected to be produced in tonnages up to 900,000 tonnes per annum.

Bollnäs

During spring 2008 Bollnäs Ovanåkers Renhållnings AB (BORAB) in Sweden, started their new waste-to-energy plant. Municipal solid waste as well as industrial waste is turned into refuse-derived fuel. The 70,000-80,000 tonnes RDF that is produced per annum is used to power the nearby BFB-plant, which provides the citizens of Bollnäs with electricity and district heating.

References

- "Biogas Flows Through Germany's Grid Big Time - Renewable Energy News Article". 14 March 2012. Archived from the original on 14 March 2012. Retrieved 17 June 2016.
- "404 - Seite nicht gefunden auf Server der Fachagentur Nachwachsende Rohstoffe e.V.: FNR" (PDF). Retrieved

17 June 2016.

- Roubík, Hynek; Mazancová, Jana; Banout, Jan; Verner, Vladimír (2016-01-20). "Addressing problems at small-scale biogas plants: a case study from central Vietnam". Journal of Cleaner Production. 112, Part 4: 2784–2792. doi:10.1016/j.jclepro.2015.09.114.

- "Biogas plants provide cooking and fertiliser". Ashden Awards, sustainable and renewable energy in the UK and developing world. Retrieved 15 May 2015.

- Administrator. "Biogas CHP - Alfagy - Profitable Greener Energy via CHP, Cogen and Biomass Boiler using Wood, Biogas, Natural Gas, Biodiesel, Vegetable Oil, Syngas and Straw". Retrieved 15 May 2015.

- Overview of Greenhouse Gases, Methane Emissions. Climate Change, United States Environmental Protection Agency, 11 December 2015.

- Pribut, Dr. Stephen. "A Brief History Of Sneakers". Dr. Stephen M. Pribut's Sport Pages. APMA NEWS. Retrieved 26 November 2014.

- Umair Irfan (April 5, 2013). "Study finds improved cookstoves solve one emissions problem, but create another". ClimateWire E & E Publishing. Retrieved April 5, 2013.

- Katarzyna Leja, Grazyna Lewandowicz. "Polymer Biodegradation and Biodegradable Polymers-a Review." Polish J. of Environmental Studies 2nd ser. 19.2010 (2012): 255-66. Web.

- "Egyptian cities and markets: What's behind a name? - Street Smart - Folk - Ahram Online". English.ahram.org.eg. 2012-06-28. Retrieved 2012-10-11.

- Miller, Naomi (1984-01-01). "The use of dung as fuel: an ethnographic example and an archaeological application | Naomi Miller". Academia.edu. Retrieved 2012-10-11.

- Wieland, P. 20128/pdf "Biomass Digestion in Agriculture: A Successful Pathway for the Energy Production and Waste Treatment in Germany" Check |url= value (help). Engineering in Life Science. Retrieved 5 November 2011.

- "Erneuerbare Energien in Deutschland. Rückblick und Stand des Innovationsgeschehens" (PDF). IfnE et al. Retrieved 5 November 2011.

- National Non-Food Crops Centre. "NNFCC Renewable Fuels and Energy Factsheet: Anaerobic Digestion", Retrieved on 2011-02-16

- "Cold climates no bar to biogas production". New Scientist. London: Sunita Harrington. 6 November 2010. p. 14. Retrieved 4 February 2011.

- Obrecht, Matevz; Denac, Matjaz (2011). "Biogas - a sustainable energy source: new measures and possibilities for SLovenia" (PDF). Journal of Energy Technology (5): 11–24.

- Wieland, P. "Production and Energetic Use of Biogas from Energy Crops and Wastes in Germany" (PDF). Applied Biochemistry and Biotechnology. Retrieved 5 November 2011.

Persistent Organic Pollutant: An Integrated Study

Persistent organic pollutants (POPs) are substances like insecticides and pesticides that do not decompose easily or are not naturally eliminated. The major categories of persistent organic pollutants are dealt in great detail. The aspects elucidated are of vital importance, and provide a better understanding of persistent organic pollutants.

Persistent Organic Pollutant

Persistent organic pollutants (POPs) are organic compounds that are resistant to environmental degradation through chemical, biological, and photolytic processes. Because of their persistence, POPs bioaccumulate with potential significant impacts on human health and the environment. The effect of POPs on human and environmental health was discussed, with intention to eliminate or severely restrict their production, by the international community at the Stockholm Convention on Persistent Organic Pollutants in 2001.

Many POPs are currently or were in the past used as pesticides, solvents, pharmaceuticals, and industrial chemicals. Although some POPs arise naturally, for example volcanoes and various biosynthetic pathways, most are man-made via total synthesis.

Consequences of Persistence

POPs typically are halogenated organic compounds and as such exhibit high lipid solubility. For this reason, they bioaccumulate in fatty tissues. Halogenated compounds also exhibit great stability reflecting the nonreactivity of C-Cl bonds toward hydrolysis and photolytic degradation. The stability and lipophilicity of organic compounds often correlates with their halogen content, thus polyhalogenated organic compounds are of particular concern. They exert their negative effects on the environment through two processes, long range transport, which allows them to travel far from their source, and bioaccumulation, which reconcentrates these chemical compounds to potentially dangerous levels. Compounds that make up POPs are also classed as PBTs (Persistent, Bioaccumulative and Toxic) or TOMPs (Toxic Organic Micro Pollutants).

Long-Range Transport

POPs enter the gas phase under certain environmental temperatures and volatize from soils, vegetation, and bodies of water into the atmosphere, resisting breakdown reactions in the air, to travel long distances before being re-deposited. This results in accumulation of POPs in areas far from where they were used or emitted, specifically environments where POPs have never been introduced such as Antarctica, and the Arctic circle. POPs can be present as vapors in the atmosphere or bound to the surface of solid particles. POPs have low solubility in water but are easily captured by solid particles, and are soluble in organic fluids (oils, fats, and liquid fuels). POPs are not easily degraded in the environment due to their stability and low decomposition rates. Due to this capacity for long-range transport, POP environmental contamination is extensive, even in areas where POPs have never been used, and will remain in these environments years after restrictions implemented due to their resistance to degradation.

Bioaccumulation

Bioaccumulation of POPs is typically associated with the compounds high lipid solubility and ability to accumulate in the fatty tissues of living organisms for long periods of time. Persistent chemicals tend to have higher concentrations and are eliminated more slowly. Dietary accumulation or bioaccumulation is another hallmark characteristic of POPs, as POPs move up the food chain, they increase in concentration as they are processed and metabolized in certain tissues of organisms. The natural capacity for animals gastrointestinal tract concentrate ingested chemicals, along with poorly metabolized and hydrophobic nature of POPs makes such compounds highly susceptible to bioaccumulation. Thus POPs not only persist in the environment, but also as they are taken in by animals they bioaccumulate, increasing their concentration and toxicity in the environment.

Stockholm Convention on Persistent Organic Pollutants

State parties to the Stockholm Convention on Persistent Organic Pollutants

The Stockholm Convention was adopted and put into practice by the United Nations Environment Programme (UNEP) on May 22, 2001. The UNEP decided that POP regulation needed to be addressed globally for the future. The purpose statement of the agreement is "to protect human health and the environment from persistent organic pollutants." As of 2014, there are 179 countries in compliance with the Stockholm convention. The convention and its participants have recognized the potential human and environmental toxicity of POPs. They recognize that POPs have the potential for long range transport and bioaccumulation and biomagnification. The convention seeks to study and then judge whether or not a number of chemicals that have been developed with advances in technology and science can be categorized as POPs or not. The initial meeting in 2001 made a preliminary list, termed the "dirty dozen," of chemicals that are classified as POPs. As of 2014, the United States of America has signed the Stockholm Convention but has not ratified it. There are a handful of other countries that have not ratified the convention but most countries in the world have ratified the convention.

Compounds on The Stockholm Convention List

In May 1995, the United Nations Environment Programme Governing Council investigated POPs. Initially the Convention recognized only twelve POPs for their adverse effects on human health and the environment, placing a global ban on these particularly harmful and toxic compounds and requiring its parties to take measures to eliminate or reduce the release of POPs in the environment.

1. Aldrin, an insecticide used in soils to kill termites, grasshoppers, Western corn rootworm, and others, is also known to kill birds, fish, and humans. Humans are primarily exposed to aldrin through dairy products and animal meats.

2. Chlordane, an insecticide used to control termites and on a range of agricultural crops, is known to be lethal in various species of birds, including mallard ducks, bobwhite quail, and pink shrimp; it is a chemical that remains in the soil with a reported half-life of one year. Chlordane has been postulated to affect the human immune system and is classified as a possible human carcinogen. Chlordane air pollution is believed the primary route of humane exposure.

3. Dieldrin, a pesticide used to control termites, textile pests, insect-borne diseases and insects living in agricultural soils. In soil and insects, aldrin can be oxidized, resulting in rapid conversion to dieldrin. Dieldrin's half-life is approximately five years. Dieldrin is highly toxic to fish and other aquatic animals, particularly frogs, whose embryos can develop spinal deformities after exposure to low levels. Dieldrin has been linked to Parkinson's disease, breast cancer, and classified as immunotoxic, neurotoxic, with endocrine disrupting capacity. Dieldrin residues have been found in air, water, soil, fish, birds, and mammals. Human exposure to dieldrin primarily derives from food.

4. Endrin, an insecticide sprayed on the leaves of crops, and used to control rodents. Animals can metabolize endrin, so fatty tissue accumulation is not an issue, however the chemical has a long half-life in soil for up to 12 years. Endrin is highly toxic to aquatic animals and humans as a neurotoxin. Human exposure results primarily through food.

5. Heptachlor, a pesticide primarily used to kill soil insects and termites, along with cotton insects, grasshoppers, other crop pests, and malaria-carrying mosquitoes. Heptachlor, even at every low doses has been associated with the decline of several wild bird populations – Canada geese and American kestrels. In laboratory tests have shown high-dose heptachlor as lethal, with adverse behavioral changes and reduced reproductive success at low-doses, and is classified as a possible human carcinogen. Human exposure primarily results from food.

6. Hexachlorobenzene (HCB), was first introduced in 1945–1959 to treat seeds because it can kill fungi on food crops. HCB-treated seed grain consumption is associated with photosensitive skin lesions, colic, debilitation, and a metabolic disorder called porphyria turcica, which can be lethal. Mothers who pass HCB to their infants through the placenta and breast milk had limited reproductive success including infant death. Human exposure is primarily from food.

7. Mirex, an insecticide used against ants and termites or as a flame retardant in plastics, rubber, and electrical goods. Mirex is one of the most stable and persistent pesticides, with a half-life of up to 10 years. Mirex is toxic to several plant, fish and crustacean species, with suggested carcinogenic capacity in humans. Humans are exposed primarily through animal meat, fish, and wild game.

8. Toxaphene, an insecticide used on cotton, cereal, grain, fruits, nuts, and vegetables, as well as for tick and mite control in livestock. Widespread toxaphene use in the US and chemical persistence, with a half-life of up to 12 years in soil, results in residual toxaphene in the environment. Toxaphene is highly toxic to fish, inducing dramatic weight loss and reduced egg viability. Human exposure primarily results from food. While human toxicity to direct toxaphene exposure is low, the compound is classified as a possible human carcinogen.

9. Polychlorinated biphenyls (PCBs), used as heat exchange fluids, in electrical transformers, and capacitors, and as additives in paint, carbonless copy paper, and plastics. Persistence varies with degree of halogenation, an estimated half-life of 10 years. PCBs are toxic to fish at high doses, and associated with spawning failure at low doses. Human exposure occurs through food, and is associated with reproductive failure and immune suppression. Immediate effects of PCB exposure include pigmentation of nails and mucous membranes and swelling of the eyelids, along with fatigue, nausea, and vomiting. Effects are transgenerational, as the chemical can persist in a mother's body for up to 7 years, resulting in developmental delays and behavioral problems in her children. Food contamination has led to large scale PCB exposure.

10. Dichlorodiphenyltrichloroethane (DDT) is probably the most infamous POP. It was widely used as insecticide during WWII to protect against malaria and typhus. After the war, DDT was used as an agricultural insecticide. In 1962, the American biologist Rachel Carson published Silent Spring, describing the impact of DDT spraying on the US environment and human health. DDT's persistence in the soil for up to 10–15 years after application has resulted in widespread and persistent DDT residues throughout the world including the arctic, even though it has been banned or severely restricted in most of the world. DDT is toxic to many organisms including birds where it is detrimental to reproduction due to eggshell thinning. DDT can be detected in foods from all over the world and food-borne DDT remains the greatest source of human exposure. Short-term acute effects of DDT on humans are limited, however long-term exposure has been associated with chronic health effects including increased risk of cancer and diabetes, reduced reproductive success, and neurological disease.

11. Dioxins are unintentional by-products of high-temperature processes, such as incomplete combustion and pesticide production. Dioxins are typically emitted from the burning of hospital waste, municipal waste, and hazardous waste, along with automobile emissions, peat, coal, and wood. Dioxins have been associated with several adverse effects in humans, including immune and enzyme disorders, chloracne, and are classified as a possible human carcinogen. In laboratory studies of dioxin effects an increase in birth defects and stillbirths, and lethal exposure have been associated with the substances. Food, particularly from animals, is the principal source of human exposure to dioxins.

12. Polychlorinated dibenzofurans are by-products of high-temperature processes, such as incomplete combustion after waste incineration or in automobiles, pesticide production, and polychlorinated biphenyl production. Structurally similar to dioxins, the two compounds share toxic effects. Furans persist in the environment and classified as possible human carcinogens. Human exposure to furans primarily results from food, particularly animal products.

New POPs on the Stockholm Convention List

Since 2001, this list has been expanded to include some polycyclic aromatic hydrocarbons (PAHs), brominated flame retardants, and other compounds. Additions to the initial 2001 Stockholm Convention list are as following POPs:

- Chlordecone, a synthetic chlorinated organic compound,is primarily used as an agricultural pesticide, related to DDT and Mirex. Chlordecone is toxic to aquatic organisms, and classified as a possible human carcinogen. Many countries have banned chlordecone sale and use, or intend to phase out stockpiles and wastes.

- α-Hexachlorocyclohexane (α-HCH) and β-Hexachlorocyclohexane (β-HCH) are insecticides as well as by-products in the production of lindane. Large stockpiles of HCH isomers exist in the environment. α-HCH and β-HCH are highly persistent in the water of colder regions. α-HCH and β-HCH has been linked Parkinson's and Alzheimer's disease.

- Hexabromodiphenyl ether (hexaBDE) and heptabromodiphenyl ether (heptaBDE) are main components of commercial octabromodiphenyl ether (octaBDE). Commercial octaBDE is highly persistent in the environment, whose only degradation pathway is through debromination and the production of bromodiphenyl ethers, which can increase toxicity.

- Lindane (γ-hexachlorocyclohexane), a pesticide used as a broad spectrum insecticide for seed, soil, leaf, tree and wood treatment, and against ectoparasites in animals and humans (head lice and scabies). Lindane rapidly bioconcentrates. It is immunotoxic, neurotoxic, carcinogenic, linked to liver and kidney damage as well as adverse reproductive and developmental effects in laboratory animals and aquatic organisms. Production of lindane unintentionally produces two other POPs α-HCH and β-HCH.

- Pentachlorobenzene (PeCB), is a pesticide and unintentional byproduct. PeCB has also been used in PCB products, dyestuff carriers, as a fungicide, a flame retardant, and a chemical intermediate. PeCB is moderately toxic to humane, while highly toxic to aquatic organisms.

- Tetrabromodiphenyl ether (tetraBDE) and pentabromodiphenyl ether (pentaBDE) are industrial chemicals and the main components of commercial pentabromodiphenyl ether (pentaBDE). PentaBDE has been detected in humans in all regions of the world.

- Perfluorooctanesulfonic acid (PFOS) and its salts are used in the production of fluoropolymers. PFOS and related compounds are extremely persistent, bioaccumulating and biomagnifying. The negative effects of trace levels of PFOS have not been established.

- Endosulfans are insecticides to control pests on crops such coffee, cotton, rice and sorghum and soybeans, tsetse flies, ectoparasites of cattle. They are used as a wood preservative. Global use and manufacturing of endosulfan has been banned under the Stockholm convention in 2011, although many countries had previously banned or introduced phase-outs of the chemical when the ban was announced. Toxic to humans and aquatic and terrestrial organisms, linked to congenital physical disorders, mental retardation, and death. Endosulfans' negative health effects are primarily liked to its endocrine disrupting capacity acting as an antiandrogen.

- Hexabromocyclododecane (HBCD) is a brominated flame retardant primarily used in thermal insulation in the building industry. HBCD is persistent, toxic and ecotoxic, with bioaccumulative and long-range transport properties.

Additive and Synergistic Effects

Evaluation of the effects of POPs on health is very challenging in the laboratory setting. For example, for organisms exposed to a mixture of POPs, the effects are assumed to be additive. Mixtures of POPs can in principle produce synergistic effects. With synergistic effects, the toxicity of each compound is enhanced (or depressed) by the presence of other compounds in the mixture. When put together, the effects can far exceed the approximated additive effects of the POP compound mixture.

Health Effects

POP exposure may cause developmental defects, chronic illnesses, and death. Some are carcinogens per IARC, possibly including breast cancer. Many POPs are capable of endocrine disruption within the reproductive system, the central nervous system, or the immune system. People and animals are exposed to POPs mostly through their diet, occupationally, or while growing in the womb. For humans not exposed to POPs through accidental or occupational means, over 90% of exposure comes from animal product foods due to bioaccumulation in fat tissues and bioaccumulate through the food chain. In general, POP serum levels increase with age and tend to be higher in females than males.

Studies have investigated the correlation between low level exposure of POPs and various diseases. In order to assess disease risk due to POPs in a particular location, government agencies may produce a human health risk assessment which takes into account the pollutants' bioavailability and their dose-response relationships.

Endocrine Disruption

The majority of POPs are known to disrupt normal functioning of the endocrine system, for example all of the dirty dozen are endocrine disruptors. Low level exposure to POPs during critical developmental periods of fetus, newborn and child can have a lasting effect throughout its lifespan. A 2002 study synthesizes data on endocrine disruption and health complications from exposure to POPs during critical developmental stages in an organism's lifespan. The study aimed to answer the question whether or not chronic, low level exposure to POPs can have a health impact on the endocrine system and development of organisms from different species. The study found that exposure of POPs during a critical developmental time frame can produce a permanent changes in the organisms path of development. Exposure of POPs during non-critical developmental time frames may not lead to detectable diseases and health complications later in their life. In wildlife, the critical development time frames are in utero, in ovo, and during reproductive periods. In humans, the critical development timeframe is during fetal development.

Reproductive System

The same study in 2002 with evidence of a link from POPs to endocrine disruption also linked low dose exposure of POPs to reproductive health effects. The study stated that POP exposure can lead to negative health effects especially in the male reproductive system, such as decreased sperm quality and quantity, altered sex ratio and early puberty onset. For females exposed to POPs, altered reproductive tissues and pregnancy outcomes as well as endometriosis have been reported.

Exposure During Pregnancy

POP exposure during pregnancy is of particular concern to the developing fetus.

Transport Across the Placenta

A study about the transfer of POPs (14 organochlorine pesticides, 7 polychlorinated biphenyls and 14 polybrominated diphenyl ethers (PBDEs)) from Spanish mothers to their unborn fetus found that POP concentrations in serum from the mother were higher than from the umbilical cord and 50 placentas. Because transfer of the POPs from mother to fetus did not correspond with passive lipid-associated diffusion, authors suggested that POPs are actively transported across the placenta.

Gestational Weight Gain and Newborn Head Circumference

A Greek study from 2014 investigated the link between maternal weight gain during pregnancy, their PCB-exposure level and PCB level in their newborn infants, their birth weight, gestational age, and head circumference. The birth weight and head circumference of the infants was the lower, the higher POP levels during prenatal development had been, but only if mothers had either excessive or inadequate weight gain during pregnancy. No correlation between POP exposure and gestational age was found. A 2013 case-control study conducted 2009 in Indian mothers and their offspring showed prenatal exposure of two types of organochlorine pesticides (HCH, DDT and DDE) impaired the growth of the fetus, reduced the birth weight, length, head circumference and chest circumference.

Cardiovascular Disease and Cancer

POPs are lipophilic environmental toxins. They are often found in lipoproteins of organisms. A study published in 2014 found an association between the concentration of POPs in lipoproteins and the occurrence of cardiovascular disease and various cancers in human beings. The higher the concentration of POPs found in lipoproteins, the higher the occurrence of cardiovascular disease and cancer. Highly chlorinated polychlorinated biphenyls are specifically found in high concentrations in lipoproteins. Cardiovascular disease is shown to be more associated with higher concentrations of POPs in high density lipoproteins and cancer is shown to be more associated with higher concentrations of POPs in low density lipoproteins and very low density lipoproteins.

Obesity

There have been many recent studies assessing the connection between serum POP levels in individuals and instances of obesity. A study released in 2011 found correlations between different POPs and obesity occurrence in individuals tested. The statistically significant findings from the study show that there is actually a negative correlation between various PCB congener serum levels and obesity in individuals tested. The study also showed a positive correlation between beta-hexachlorocyclohexane and various dioxin serum levels and obesity in individuals tested. Obesity was determined using the Body Mass Index (BMI). One proposed explanation in the study is that PCBs are very lipophilic, therefore they are easily stored and captured in the fat deposits in human beings. Obese individuals have higher amounts of fat deposits in their body, and thus more PCBs could be captured in the fat deposits leading to less PCBs circulating in blood serum. The study provides evidence demonstrating that the correlation between POP serum levels and obesity occurrence is more complicated than previously expected. The same study also noted a strong positive correlation between serum POP levels and age in all individuals in the experiment.

Diabetes

A study published in 2006 revealed a positive correlation between POP serum levels and type II diabetes in individuals, after other variables, such as age, sex, race, and socioeconomic status were adjusted for. The correlation proved stronger in younger, Mexican American, and obese individuals. Individuals exposed to low doses of POPs throughout their lifetime had a higher chance for developing diabetes than individuals exposed to high concentrations of POPs for a short amount of time.

POPs in Urban Areas and Indoor Environments

Traditionally it was thought that human exposure to POPs occurred primarily through food, however indoor pollution patterns that characterize certain POPs have challenged this notion. Recent studies of indoor dust and air have implicated indoor environments as a major sources for human exposure via inhalation and ingestion. Furthermore, significant indoor POP pollution must be a major route of human POP exposure, considering the modern trend in spending larger proportions of life indoor. Several studies have shown that indoor (air and dust) POP levels to exceed outdoor (air and soil) POP concentrations.

Control and Removal of POPs in the Environment

Current studies aimed at minimizing POPs in the environment are investigating their behavior in photo catalytic oxidation reactions. POPs that are found in humans and in aquatic environments the most are the main subjects of these experiments. Aromatic and aliphatic degradation products have been identified in these reactions. Photochemical degradation is negligible compared to photocatalytic degradation. However, proper removal techniques of POPs from the environment are still unclear, due to fear that more toxic byproducts may result from uninvestigated degradation techniques. Current efforts are more focused on banning the use and production of POPs worldwide rather than removal of POPs.

Persistent, Bioaccumulative and Toxic Substances

Persistent, bioaccumulative and toxic (PBTs) substances are a class of compounds that have high resistance to degradation from abiotic and biotic factors, high mobility in the environment and high toxicity. Because of these factors PBTs have been observed to have a high order of bioaccumulation and biomagnification, very long retention times in various media, and widespread distribution across the globe. Majority of PBTs in the environment are either created through industry or are unintentional byproducts.

History

Persistent organic pollutants (POPs) were the focal point of the Stockholm Convention 2001 due to their persistence, ability to biomagnify and the threat posed to both human health and the environment. The goal of the Stockholm Convention was to determine the classification of POPs, create measures to eliminate production/use of POPs, and establish proper disposal of the compounds in an environmentally friendly manner. Currently the majority of the global community is actively involved with this program but a few still resist, most notably the U.S.

Similar to POPs classification, the PBT classification of chemicals was developed in 1997 by the Great Lakes Binational Toxic Strategy (GLBNS). Signed by both the U.S and Canada, the GLBNS classified PBTs in one of two categories, level I and level II. Level I PBTs are top priority which currently, as of 2005, contained 12 compounds or classes of compounds.

Level I PBTs (GLBNS)

- Mercury

- Polychlorinated biphenyls PCBs)

- Dioxins/furans

- Benzo(a)pyrene (BaP)

- Hexachlorobenzene (HCB)

- Alkyl-lead

- Pesticides

    o Mirex

    o Dieldrin/aldrin

    o Chlordane

    o Toxaphene

- Octachlorostyrene

The GLBNS is administered by the U.S Environmental Protection Agency (USEPA) and Environment Canada. Following the GLBNS, the Multimedia Strategy for Priority Persistent, Bioaccumulative and Toxic Pollutants (PBT Strategy) was drafted by the USEPA. The PBT Strategy led to the implementation of PBT criteria in several regulational policies. Two main policies that were changed by the PBT strategy were the Toxics Release Inventory (TRI) which required more rigid chemical reporting and the New Chemical Program (NCP) under the Toxics Substances Control Act (TSCA) which required screening for PBTs and PBT properties.

Compounds

General

PBTs are a unique classification of chemicals that have and will continue to impact human health and the environment worldwide. The three main attributes of a PBT (persistence, bioaccumulative and toxic) each have a huge role in the risk posed by these compounds.

Persistence

PBTs have a high environmental mobility relative to other contaminants mainly due to their resistance to degradation (persistence). This allows PBTs to travel far and wide in both the atmosphere and in aqueous environments. The low degradation rates of PBTs allow these chemicals to be exposed to both biotic and abiotic factors while maintaining a relatively stable concentration. Another factor that makes PBTs especially dangerous are the degradation products which are often relatively as toxic as the parent compound. These factors have resulted in global contamination most notable in remote areas such as the arctic and high elevation areas which are far from any source of PBTs.

Bioaccumulation and Biomagnification

The bioaccumulative ability of PBTs follows suit with the persistence attribute by the high resistance to degradation by biotic factors, especially with in organisms. Bioaccumulation is the result of a toxic substance being taken up at a higher rate than being removed from an organism. For PBTs this is caused mainly by a resistance to degradation, biotic and abiotic. PBTs usually are highly insoluble in water which allows them to enter organisms at faster rates through fats and other nonpolar regions on an organism. Bioaccumulation of a toxicant can lead to biomagnification through a trophic web which has resulted in massive concern in areas with especially low trophic diversity. Biomagnification results in higher trophic organisms accumulating more PBTs than those of lower trophic levels through consumption of the PBT contaminated lower trophic organisms.

Toxicity

The toxicity of this class of compounds is high with very low concentrations of a PBT required to enact an effect on an organism compared to most other contaminants. This high toxicity along with the persistence allows for the PBT to have detrimental effects in remote areas around the globe where there is not a local source of PBTs. The bioaccumulation and magnification along with the high toxicity and persistence has the ability to destroy and/or irreparably damage trophic systems, especially the higher trophic levels, globally. It is this reason that PBTs have become an area of focus in global politics.

Specific Toxicants

PCBs

Historically, PCBs were used extensively for industrial purposes such as coolants, insulating fluids, and as a plasticizer. These contaminants enter the environment through both use and disposal. Due to extensive concern from the public, legal, and scientific sectors indicating that PCBs are likely carcinogens and potential to adversely impact the environment, these compounds were banned in 1979 in the United States. The ban included the use of PCBs in uncontained sources, such as adhesives, fire retardant fabric treatments, and plasticizers in paints and cements. Containers that are completely enclosed such as transformers and capacitors are exempt from the ban.

The inclusion of PCBs as a PBT may be contributed to their low water solubility, high stability, and semi-volatility facilitating their long range transport and accumulation in organisms. The persistence of these compounds is due to the high resistance to oxidation, reduction, addition, elimination and electrophilic substitution. The toxicological interactions of PCBs are affected by the number and position of the chlorine atoms, without ortho substitution are referred as coplanar and all others as non-coplanar. Non-coplanar PCBs may cause neurotoxicity by interfering with intracellular signal transduction dependent on calcium. Ortho-PCBs may alter hormone regulation through disruption of the thyroid hormone transport by binding to transthyretin. Coplanar PCBs are similar to dioxins and furans, both bind to the aryl hydrocarbon receptor (AhR) in organisms and may exert dioxin-like effects, in addition to the effects shared with non-coplanar PCBs. The AhR is a transcription factor, therefore, abnormal activation may disrupt cellular function by altering gene transcription.

Effects of PBTs may include increase in disease, lesions in benthic feeders, spawning loss, change in age-structured populations of fish, and tissue contamination in fish and shellfish. Humans and other organisms, which consume shellfish and/or fish contaminated with persistent bioaccumulative pollutants, have the potential to bioaccumulate these chemicals. This may put these organisms at risk of mutagenic, teratogenic, and/or carcinogenic effects. Correlations have been found between elevated exposure to PCB mixtures and alterations in liver enzymes, hepatomegaly, and dermatological effects such as rashes have been reported.

DDT

One PBT of concern includes DDT (dichlorodiphenyltrichloroethane), an organochlorine that was widely used as an insecticide during World War II to protect soldiers from malaria carried by mosquitoes. Due to the low cost and low toxicity to mammals, the widespread use of DDT for agricultural and commercial motives started around 1940. However, the overuse of DDT lead to insect tolerance to the chemical. It was also discovered that DDT had a high toxicity to fish. DDT was banned in the US by 1973 because of building evidence that DDT's stable structure, high fat solubility, and low rate of metabolism, caused it to bioaccumulate in animals. While DDT is banned in the US, other countries such as China and Turkey still produce and use it quite regularly through Dicofol, an insecticide that has DDT as an impurity. This continued use in other parts of the world is still a global problem due to the mobility and persistence of DDT.

The initial contact from DDT is on vegetation and soil. From here, the DDT can travel many routes, for instance, when plants and vegetation are exposed to the chemical to protect from insects, the plants may absorb it. Then these plants may either be consumed by humans or other animals. These consumers ingest the chemical and begin metabolizing the toxicant, accumulating more through ingestion, and posing health risks to the organism, their offspring, and any predators. Alternatively the ingestion of the contaminated plant by insects may lead to tolerance by the organism. Another route is the chemical travelling through the soil and ending up in ground water and in human water supply. Or in the case that the soil is near a moving water system, the chemical could end up in large freshwater systems or the ocean where fish are at high risk from the toxicological effects of DDT. Lastly, the most common transport route is the evaporation of DDT into the atmosphere followed by condensation and eventually precipitation where it is released into environments anywhere on earth. Due to the long-range transport of DDT, the presence of this harmful toxicant will continue as long as it is still used anywhere and until the current contamination eventually degrades. Even after its complete discontinued use, it will still remain in the environment for many more years after because of DDT's persistent attributes.

Previous studies have shown that DDT and other similar chemicals directly elicited a response and effect from excitable membranes. DDT causes membranes such as sense organs and nerves endings to activate repetitively by slowing down the ability for the sodium channel to close and stop releasing sodium ions. The sodium ions are what polarize the opposing synapse after it has depolarized from firing. This inhibition of closing the sodium ion channel can lead to a variety of problems including a dysfunctional nervous system, decreased motor abilities/function/control, reproductive impairment (egg-shell thinning in birds), and development deficiencies. Presently, DDT has been labeled as a possible human carcinogen based on animal liver tumor studies. DDT toxicity on humans have been associated with dizziness, tremors, irritability, and convulsions. Chronic toxicity has led to long term neurological and cognitive issues.

Mercury

Inorganic

Inorganic mercury (elemental mercury) is less bioavailable and less toxic than that of organic mercury but is still toxic nonetheless. It is released into the environment through both natural sources as well as human activity, and it has the capability to travel long distances through the atmosphere. Around 2,700 to 6,000 tons of elemental mercury are released via natural activity such as volcanoes and erosion. Another 2,000 – 3,000 tons are released by human industrial activities such as coal combustion, metal smelting and cement production. Due to its high volatility and atmospheric residence time of around 1 year, mercury has the ability to travel across continents before being deposited. Inorganic mercury has a wide spectrum of toxicological effects that include damage to the respiratory, nervous, immune and excretory systems in humans. Inorganic mercury also possesses the ability to bioaccumulate individuals and biomagnify through trophic systems.

Organic

Organic mercury is significantly more detrimental to the environment than its inorganic form due to its widespread distribution as well as its higher mobility, general toxicity and rates of bioaccumulation than that of the inorganic form. Environmental organic mercury is mainly created by the transformation of elemental (inorganic) mercury via anaerobic bacteria into methylated mercury (organic). The global distribution of organic mercury is the result of general mobility of the compound, activation via bacteria and transportation from animal consumption. Organic mercury shares a lot of the same effects as the inorganic form but it has a higher toxicity due to its higher mobility in the body, especially its ability to readily move across the blood brain barrier.

Ecological Impact of Hg

The high toxicity of both forms of mercury (especially organic mercury) poses a threat to almost all organisms that comes in contact with it. This is one of the reasons that there is such high attention to mercury in the environment but even more so than its toxicity is both its persistence and atmospheric retention times. The ability of mercury to readily volatilize allows it to enter the atmosphere and travel far and wide. Unlike most other PBTs that have atmospheric half-lives between 30 min and 7 days mercury has an atmospheric residence time of at least 1 year. This atmospheric retention time along with mercury's resistance to degradation factors such as electromagnetic radiation and oxidation, which are two of the main factors leading to degradation of many PBTs in the atmosphere, allows mercury from any source to be transported extensively. This characteristic of mercury transportation globally along with its high toxicity is the reasoning behind its incorporation into the BNS list of PBTs.

Notable PBT environmental impacts

Japan

The realization of the adverse effects from environmental pollution were made public from several disasters that occurred globally. In 1965, it was recognized that extensive mercury pollution by the Chisso chemical factory in Minamata, Japan due to improper handling of industrial wastes resulted in significant effects to the humans and organisms exposed. Mercury was released into the environment as methyl mercury (bioavailable state) into industrial wastewater and was then bioaccumulated by shellfish and fish in Minamata Bay and the Shiranui Sea. When the contaminated seafood was consumed by the local populace it caused a neurological syndrome, coined Minamata disease. Symptoms include general muscle weakness, hearing damage, reduced field of vision, and ataxia. The Minamata disaster contributed to the global realization of the potential dangers from environmental pollution and to the characterization of PBTs.

Puget Sound

Despite the ban on DDT 30 years earlier and years of various efforts to clean up Puget Sound from DDT and PCB's, there is still a significant presence of both compounds which pose a constant threat to human health and the environment. Harbor seals (*Phoca vitulina*), a common marine species in the Puget Sound area have been the focus of a few studies to monitor and examine the effects of DDT accumulation and magnification in aquatic wildlife. One study tagged and reexamined seal pups every 4 to 5 years to be tested for DDT concentrations. The trends showed the pups to be highly contaminated; this means their prey are also highly contaminated. Due to DDT's high lipid solubility, it also has the ability to accumulate in the local populace who consume seafood from the area. This also translates to women who are pregnant or breastfeeding, since DDT will be transferred from the mother to child. Both animal and human health risk to DDT will continue to be an issue in Puget Sound especially because of the cultural significance of fish in this region.

Stockholm Convention on Persistent Organic Pollutants

State parties to the Stockholm Convention as of May 2012

Stockholm Convention on Persistent Organic Pollutants is an international environmental treaty, signed in 2001 and effective from May 2004, that aims to eliminate or restrict the production and use of persistent organic pollutants (POPs).

History

In 1995, the Governing Council of the United Nations Environment Programme (UNEP) called for global action to be taken on POPs, which it defined as "chemical substances that persist in the environment, bio-accumulate through the food web, and pose a risk of causing adverse effects to human health and the environment".

Following this, the Intergovernmental Forum on Chemical Safety (IFCS) and the International Programme on Chemical Safety (IPCS) prepared an assessment of the 12 worst offenders, known as the *dirty dozen*.

The INC met five times between June 1998 and December 2000 to elaborate the convention, and delegates adopted the Stockholm Convention on POPs at the Conference of the Plenipotentiaries convened from 22–23 May 2001 in Stockholm, Sweden.

The negotiations for the Convention were completed on 23 May 2001 in Stockholm. The convention entered into force on 17 May 2004 with ratification by an initial 128 parties and 151 signatories. Co-signatories agree to outlaw nine of the dirty dozen chemicals, limit the use of DDT to malaria control, and curtail inadvertent production of dioxins and furans.

Parties to the convention have agreed to a process by which persistent toxic compounds can be reviewed and added to the convention, if they meet certain criteria for persistence and transboundary threat. The first set of new chemicals to be added to the Convention were agreed at a conference in Geneva on 8 May 2009.

As of March 2016, there are 180 parties to the Convention, (179 states and the European Union). Notable non-ratifying states include the United States, Israel, Malaysia, and Italy.

The Stockholm Convention was adopted to EU legislation in REGULATION (EC) No 850/2004.

Summary of Provisions

Key elements of the Convention include the requirement that developed countries provide new and additional financial resources and measures to eliminate production and use of intentionally produced POPs, eliminate unintentionally produced POPs where feasible, and manage and dispose of POPs wastes in an environmentally sound manner. Precaution is exercised throughout the Stockholm Convention, with specific references in the preamble, the objective, and the provision on identifying new POPs.

Persistent Organic Pollutants Review Committee

When adopting the Convention, provision was made for a procedure to identify additional POPs and the criteria to be considered in doing so. At the first meeting of the Conference of the Parties (COP1), held in Punta del Este, Uruguay from 2–6 May 2005, the POPRC was established to consider additional candidates nominated for listing under the Convention.

The Committee is composed of 31 experts nominated by parties from the five United Nations regional groups and reviews nominated chemicals in three stages. The Committee first determines whether the substance fulfills POP screening criteria detailed in Annex D of the Convention, relating to its persistence, bioaccumulation, potential for long-range environmental transport (LRET), and toxicity. If a substance is deemed to fulfill these requirements, the Committee then drafts a risk profile according to Annex E to evaluate whether the substance is likely, as a result of its LRET, to lead to significant adverse human health and/or environmental effects and therefore warrants global action. Finally, if the POPRC finds that global action is warranted, it develops a risk management evaluation, according to Annex F, reflecting socioeconomic considerations associated with possible control measures. Based on this, the POPRC decides to recommend that the COP list the substance under one or more of the annexes to the Convention. The POPRC has met annually in Geneva, Switzerland since its establishment.

The seventh meeting of the Persistent Organic Pollutants Review Committee (POPRC-7) of the Stockholm Convention on Persistent Organic Pollutants (POPs) took place from 10–14 October 2011 in Geneva, Switzerland. POPRC-8 was held from 15–19 October 2012 in Geneva, and POPRC-9 was held from 14–18 October 2013 in Rome.

Listed Substances

There were initially twelve distinct chemicals listed in three categories. Two chemicals, hexachlorobenzene and polychlorinated biphenyls, were listed in both categories A and C.

Annex	Name	CAS Number	Exemptions
A. Elimination	Aldrin	309-00-2	Production none Use as a local ectoparasiticide and insecticide
A. Elimination	Chlordane	57-74-9	Production by registered parties Use as a local ectoparasiticide, insecticide, termiticide (including in buildings, dams and roads) and as an additive in plywood adhesives
A. Elimination	Dieldrin	60-57-1	Production none Use in agricultural operations

Annex	Name	CAS Number	Exemptions
A. Elimination	Endrin	72-20-8	None
A. Elimination	Heptachlor	76-44-8	Production none Use as a termiticide (including in the structure of houses and underground), for organic treatment and in underground cable boxes
A. Elimination	Hexachlorobenzene	118-74-1	Production by registered parties Use as a chemical intermediate and a solvent for pesticides
A. Elimination	Mirex	2385-85-5	Production by registered parties Use as a termiticide
A. Elimination	Toxaphene	8001-35-2	None
A. Elimination	Polychlorinated biphenyls (PCBs)	various	Production none Use in accordance with part II of Annex A
B. Restriction	DDT	50-29-3	Disease vector control in accordance with Part II of Annex B Production and use as an intermediate in the production of dicofol and other compounds
C. Unintentional Production	Polychlorinated dibenzo-p-dioxins ("dioxins") and polychlorinated dibenzofurans	various	
C. Unintentional Production	Polychlorinated biphenyls (PCBs)	various	
C. Unintentional Production	Hexachlorobenzene	118-74-1	

Added by the Fourth Conference of Parties, May 2009

These modifications has come into force on 26 August 2010, except for countries that submit a notification pursuant to the provisions of paragraph 3(b) of Article 22.

Annex	Name	CAS Number	Exemptions
A. Elimination	α-Hexachlorocyclohexane	319-84-6	None
A. Elimination	β-Hexachlorocyclohexane	319-85-7	None
A. Elimination	Chlordecone	143-50-0	None
A. Elimination	Hexabromobiphenyl	36355-01-8	None
A. Elimination	Hexabromodiphenyl ether and heptabromodiphenyl ether	various	Production none Use recycling and reuse of articles containing these compounds
A. Elimination	Lindane (gamma-hexachlorocyclohexane)	58-89-9	Production none Use Human health pharmaceutical for control of head lice and scabies as second line treatment
A. Elimination & C. Unintentional Production	Pentachlorobenzene	608-93-5	None

A. Elimination	Tetrabromodiphenyl ether and pentabromodiphenyl ether	various	Production none Use recycling and reuse of articles containing these compounds
B. Restriction	Perfluorooctanesulfonic acid (PFOS), its salts and perfluorooctanesulfonyl fluoride (PFOSF)	various	Production for permitted uses Use various uses specified in part III of Annex B

Added by the Fifth Conference of Parties, May 2011

Annex	Name	CAS Number	Exemptions
A. Elimination	Endosulfan	115-29-7 959-98-8 33213-65-9	Production As allowed for the parties listed in the Register of specific exemptions Use Crop-pest complexes as listed in accordance with the provisions of part VI of Annex A.

Added by the Sixth Conference of Parties, April–May 2013

These modifications has come into force on 26 November 2014, except for countries that submit a notification pursuant to the provisions of paragraph 3(b) of Article 22.

Annex	Name	CAS Number	Exemptions
A. Elimination	Hexabromocyclododecane	25637-99-4 3194-55-6 134237-50-6 134237-51-7 134237-52-8	Production As allowed for the parties listed in the Register in accordance with the provisions of Part VII of this Annex Use Expanded polystyrene and extruded polystyrene in buildings in accordance with the provisions of Part VII of this Annex.

Chemicals Newly Proposed for Inclusion in Annexes A,B,C

POPRC-7 considered three proposals for listing in Annexes A, B and/or C of the Convention: chlorinated naphthalenes (CNs), hexachlorobutadiene (HCBD) and pentachlorophenol (PCP), its salts and esters. The proposal is the first stage of the POPRC's work in assessing a substance, and requires the POPRC to assess whether the proposed chemical satisfies the criteria in Annex D of the Convention. The criteria for forwarding a proposed chemical to the risk profile preparation stage are persistence, bioaccumulation, potential for long-range environmental transport (LRET), and adverse effects.

POPRC-8 proposed hexabromocyclododecane for listing in Annex A, with specific exemptions for production and use in expanded polystyrene and extruded polystyrene in buildings. This proposal was agreed at the sixth Conference of Parties on 28 April-10 May 2013.

POPRC-9 proposed di-,tri-,tetra-,penta-,hexa-, hepta- and octa-chlorinated napthalenes, and hexachlorobutadiene for listing in Annexes A and C. It also set up further work on pentachlorophenol, its salts and esters, and decabromodiphenyl ether, perfluorooctanesulfonic acid, its salts and perfluorooctane sulfonyl chloride.

Controversies

Although some critics have alleged that the treaty is responsible for the continuing death toll from malaria, in reality the treaty specifically permits the public health use of DDT for the control of mosquitoes (the malaria vector). From a developing country perspective, a lack of data and information about the sources, releases, and environmental levels of POPs hampers negotiations on specific compounds, and indicates a strong need for research.

Related conventions and other ongoing negotiations regarding pollution

- Rotterdam Convention on the Prior Informed Consent Procedure for Certain Hazardous Chemicals and Pesticides in International Trade

- Convention on Long-Range Transboundary Air Pollution (CLRTAP)

- Basel Convention on the Control of Transboundary Movements of Hazardous Wastes and their Disposal

Ongoing negotiations

- Intergovernmental Negotiating Committee's work towards a Legally Binding Instrument on Mercury

- Intergovernmental Forum on Chemical Safety (IFCS)

- Strategic Approach to International Chemicals Management (SAICM)

Aarhus Protocol on Persistent Organic Pollutants

The Aarhus Protocol on Persistent Organic Pollutants, a 1998 protocol on persistent organic pollutants (POPs), is an addition to the 1979 Geneva Convention on Long-Range Transboundary Air Pollution (LRTAP). The Aarhus POPs Protocol seeks "to control, reduce or eliminate discharge, emissions and losses of persistent organic pollutants" in Europe, some former Soviet Union countries, and the United States.

The protocol was amended on 18 December 2009, but the amended version has not yet come into force.

As of May 2013, the protocol has been ratified by 31 states and the European Union.

In the United States, the protocol is an executive agreement that does not require Senate approval. However, legislation is needed to resolve inconsistencies between provisions of the protocol and existing U.S. laws (specifically the Toxic Substances Control Act and the Federal Insecticide, Fungicide, and Rodenticide Act).

Substances

The following substances are contained in the CLRTAP POPs Protocol.

POP	Inclusion
Aldrin	Originally included
Chlordane	Originally included
Dieldrin	Originally included
Endrin	Originally included
Heptachlor	Originally included
Hexachlorobenzene	Originally included
Mirex	Originally included
Toxaphene	Originally included
PCBs	Originally included
DDT	Originally included
PCDDs/PCDFs	Originally included
Chlordecone	Originally included
Hexachlorocyclohexanes	Originally included
Hexabromobiphenyl	Originally included
PAHs	Originally included
Pentabromodiphenyl ether	Recognized
Octabromodiphenyl ether	Recognized
Pentachlorobenzene	Recognized
PFOS	Recognized
Hexachlorobutadiene	Recognized
PCNs	Recognized
SCCPs	Recognized

References

- Astoviza, Malena J. (15 April 2014). "Evaluación de la distribución de contaminantes orgánicos persistentes (COPs) en aire en la zona de la cuenca del Plata mediante muestreadores pasivos artificiales" (in Spanish): 160. Retrieved 16 April 2014.

- Vizcaino, E; Grimalt JO; Fernández-Somoano A; Tardon A (2014). "Transport of persistent organic pollutants across the human placenta". Environ Int. 65: 107–115. doi:10.1016/j.envint.2014.01.004. PMID 24486968.

- Ljunggren SA, Helmfrid I, Salihovic S, van Bavel B, Wingren G, Lindahl M, Karlsson H. Persistent organic pollutants distribution in lipoprotein fractions in relation to cardiovascular disease and cancer. Environ Int. 2014;65:93–99. doi:10.1016/j.envint.2013.12.017. PMID 24472825.

- Dewan, Jain V,; Gupta P; Banerjee BD. (February 2013). "Organochlorine pesticide residues in maternal blood, cord blood, placenta, and breastmilk and their relation to birth size". Chemosphere. 90 (5): 1704–1710. doi:10.1016/j.chemosphere.2012.09.083. PMID 23141556.

- USEPA. The Great Lakes Water Quality Agreement U.S Eighth Response to International Joint Commission. Retrieved June 6, 2012

- Szabo DT, Loccisano AE (March 30, 2012). A. Schecter, ed. "POPs and Human Health Risk Assessment". Dioxins and Persistent Organic Pollutants. John Wiley & Sons. 3rd. doi:10.1002/9781118184141.ch19.

- Dirinck, E., Jorens, P.G., Covaci, A., Geens, T., Roosens, L., Neels, H., Mertens, I., Van Gaal, L. (2011). Obesity and Persistent Organic pollutants: Possible Obesogenic Effect of Organochlorine Pesticides and Polychlorinated Biphenyls. Obesity. 19(4), 709–714.

- Yu, G.W.,, Laseter, J., Mylander, C. Persistent organic pollutants in serum and several different fat compartments in humans. J Environ Public Health. 2011;2011:417980. doi:10.1155/2011/417980. PMID 21647350.

- El-Shahawi, M.S., Hamza, A., Bashammakhb, A.S., Al-Saggaf, W.T. (2010). An overview on the accumulation, distribution, transformations, toxicity and analytical methods for the monitoring of persistent organic pollutants. Talanta. 80, 1587–1597. doi:10.1016/j.talanta.2009.09.055

- This article incorporates public domain material from the Congressional Research Service document "Report for Congress: Agriculture: A Glossary of Terms, Programs, and Laws, 2005 Edition" by Jasper Womach.

Permissions

Index